T0280145

Taniguchi Symposia on Brain Sciences No.7

TRANSMEMBRANE SIGNALING AND SENSATION

Taniguchi Symposia on Brain Sciences No. 7
PROGRAM COMMITTEE

The Taniguchi Foundation, Division of Brain Sciences
ORGANIZING COMMITTEE

Taniguchi Symposia on Brain Sciences No.7

TRANSMEMBRANE SIGNALING AND SENSATION

Edited by
Fumio Oosawa,
Tohru Yoshioka,
and Hiroshi Hayashi

JAPAN SCIENTIFIC SOCIETIES PRESS Tokyo
VNU SCIENCE PRESS BV Utrecht, The Netherlands

Supported in part by the Ministry of Education, Science and Culture under Grant-in-Aid for Publication of Scientific Research Result.

Published jointly by
JAPAN SCIENTIFIC SOCIETIES PRESS Tokyo
ISBN 4-7622-8424-6
and
VNU SCIENCE PRESS BV Utrecht, The Netherlands
ISBN 90-6764-041-7

Distributed in all areas outside Japan and Asia between Pakistan and Korea by VNU Science Press BV Utrecht, The Netherlands

Printed in Japan

PREFACE

Living cells respond to various stimuli and the environmental changes. Single cells of bacteria have chemo-, thermo- and photo-sensory systems in the membrane and send signals to the motility organ in the same cell. In multicellular organisms, these sensory systems are installed in specialized cells from which the signals are transmitted to nerve cells and, finally, to muscle cells. Those cells along the route of signal transmission have their own systems of information reception, transduction and response, and others which are not involved are also sensitive to particular stimuli and show varying responses.

Modern biology provides evidence that living cells are a complex composition of molecular machines, each responsible for an elementary function; the overall behavior of the cell is the result of the concerted work of the machines in a given circuit. Is there a common principle in the mechanism of these molecular machines for information reception, transduction and response in different cells? Is there a common principle in the design of their circuits? What are their similarities and differences and how have those machines and circuits evolved?

To approach these problems, the authors discuss various stimulus-response systems from bacteria and paramecia to insects and mammals. Each paper is concerned with a specific system in a specific cell: chemosensory systems of bacteria (Chapters 12 and 13), photosensory systems of bacteria (Chapter 14) and insects (Chapter 7), thermosensory systems of bacteria (Chapter 15) and paramecia (Chapter 11), mechanosensory systems of mammals (Chapters 8 and 9), and the stimulus-response systems in nerve cells, platelet cells and cultured cells (Chapters 1–4 and 18) and in synaptic vesicles (Chapter 19) and sarcoplasmic reticulum vesicles (Chapter 17). Subjects dealt with are the identification and characterization of receptor molecules, physical properties of channel molecules, physical and chemical changes in receptors, channels and other structures after stimuli, the molecular events in sequential response, excitation, adaptation and learning, and others. In these studies, both the traditional and the very newest techniques of electrophysiology (Chapters 10 and 16) biochemistry, genetics and ultrastructural analysis (Chapters 5 and 6) are utilized.

A typical stimulus-response system is found in the first step of chemoreception. Receptor proteins to bind specific chemicals and channel proteins to permeate specific ions are on the membrane of chemosensory cells. Information on the chemicals is transformed *via* conformational changes of the protein molecules into an electric potential difference across the membrane. There are many varieties of stimulus-response systems in living cells: in some cases, lipid molecules appear to be involved in the information reception. Not only are the physical conformations of protein or lipid molecules changed but their chemical structures are often modified in response to stimuli. These changes are not necessarily limited to the molecules on the membrane. The information is sometimes directly carried by transmitters rather than being transformed into the electric potential. In some cells, the response is spread throughout the cell and in others it is localized; it may be probabilistic rather than deterministic.

In each system, part of the circuit of the molecular machines has been successfully analyzed and the methodology and knowledge of one system is often applicable to other systems. At present, however, we are far from a complete understanding of these systems and can only speculate on the possibility of principles in the molecular mechanisms and the circuit designs. Nevertheless, these chapters document the dynamic revelations that are emerging on many different stimulus-response systems. The conformations of protein and/or lipid molecules have thermal fluctuations and their chemical structures undergo cyclic changes, even in the absence of stimuli. Stimuli change the forward and backward rates of fluctuations and cycles so that the average conformation and structure are shifted. These shifts appear in different time scales, and the physical conformation and chemical structure do not necessarily have a one-to-one correspondence.

In pointing out the diversity and what is known of the common aspects of the stimulus-response systems in living cells, the authors of this volume illustrate the value of different scientific approaches and demonstrate that future research must continue to combine physical and chemical concepts and techniques.

We are most grateful for the support of the Taniguchi Foundation and for a grant from the Ministry of Education, Science and Culture of Japan for making publication of this volume possible.

October 1984

F. Oosawa
T. Yoshioka
H. Hayashi

CONTENTS

1

DYNAMICS OF PHOSPHOLIPIDS AND CALCIUM IN SECRETORY CELLS

ATSUSHI IMAI, KOH YANO, SHIGERU NAKASHIMA, YASUHIRO ISHIZUKA, HIROAKI HATTORI, MASARU TAKAHASHI, YUKIO OKANO, AND YOSHINORI NOZAWA

Department of Biochemistry, Gifu University School of Medicine, Gifu 500, Japan

The mechanism(s) by which specific biochemical signals are transferred through membrane is of major importance in biology. Physiological messages such as peptide hormones and neurotransmitters are recognized by and bind to specific receptors on the surface of their target cell membranes. These interactions then initiate biochemical and physical changes in membranes which, in turn, allow cells to exert their various specific functions (*23*). It is now widely accepted that one of the earliest and most crucial membrane changes in the sequence of cellular activation is the turnover of membrane phospholipids, in particular, inositol phospholipids (*27*). In many secretory cells, rapid metabolism of inositol phospholipids has been observed to occur in parallel with secretory response (*7, 11, 20, 25*). On the other hand, the calcium mobilization within the cell can be responsible for initiating events which activate cell functions including those involved in secretion, contraction and a variety of biochemical events which support these changes (*6*). This occurs as a result of increased permeability of the plasma membrane to Ca^{2+} and through release of Ca^{2+} from intracellular stores. These concepts, although based on somewhat circumstantial evidence, have given rise to the hypothesis that phospholipid metabolism may be an intermediate in the Ca^{2+} regulating mechanism (*27*). Further recent support has advanced the role of phospholipid metabolism in stimulus-response coupling: (1) 1,2-diacylglycerol (DG), converted from phosphatidylinositol (PI) by the action of a PI-specific phospholipase C following stimulation, controls a calcium-dependent protein kinase (C-kinase) which is implicated in secretory processes of various cells (*33*); (2) 1,2-DG decreases the calcium concentration required for a phosphatidylcholine/phosphatidylserine mixture to undergo phase separation which subsequently affects the membrane perturbation (*28*); and (3) phosphatidic acid (PA), produced by phosphorylation of 1,2-DG, serves as an endogenous calcium ionophore in cells (*32*).

In this article, the correlation between phospholipid metabolism and Ca^{2+} mobilization and their relation to stimulus-response coupling have been investigated in human platelets, rabbit neutrophils and rat mast cells.

I. PHOSPHOLIPID HYDROLYSIS

It is well-known that rapid turnover of membrane phospholipid is observed upon platelet activation by physiological (receptor-mediated) stimuli such as thrombin and collagen (*4, 14, 31*). Two possible mechanisms which initiate this reaction are (1) the hydrolysis of phosphatidylcholine (PC) and phosphatidylethanolamine (PE) by phospholipase A_2 to release arachidonic acid from their 2-position (*5, 18, 26*), and (2) the cleavage of PI to 1,2-DG as a result of phospholipase C activation (*30*). 1,2-DG is followed by phosphorylation to PA with DG kinase, and then PA is finally resynthesized to PI *via* CDP-DG (PI-cycle). As illustrated in Fig. 1, when [^{3}H]arachidonate-labeled human platelets were exposed to thrombin, the radioactivity in 1,2-DG markedly increased within 10 sec, and then returned to the initial level by 5 min. The formed 1,2-DG appeared to be subsequently converted to PA. The rate of accumulation of 1,2-DG and PA in [^{3}H]arachidonate-labeled platelets was much greater than that in [^{3}H]glycerol-labeled platelets (*13*). This difference may reflect the preexisting DG and PA prior to activation which contain small amounts of arachidonate (*13*).

Additionally, lysoPC and lysoPE rose following the thrombin-activation (Fig. 1) but no changes were detected in the fraction of lysoPI (*13*). This can only be explained by the stimulation of phospholipase A_2 activity against PC and PE in thrombin-activated platelets. Billah and Lapetina (*2*) observed production of lysoPI in thrombin-activated horse platelets, suggesting the hydrolyzing activity of phospholipase A_2 on PI.

Rapid metabolisms of membrane phospholipids are also observed

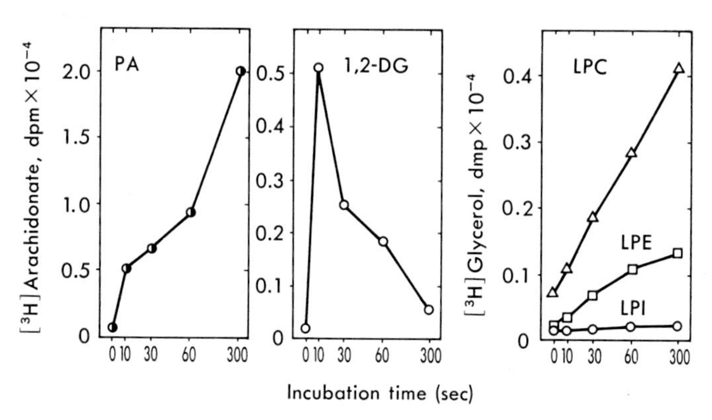

Fig. 1. Thrombin-induced phospholipid metabolism in human platelets. Washed [^{3}H]-arachidonic acid- or [^{3}H]glycerol-labeled platelets (10^9 cells) were incubated with thrombin (2 units) at 37°C for the time indicated. PA: phosphatidic acid, 1,2-DG; 1,2-diacylglycerol; LPC: lysophosphatidylcholine; LPE: lysophosphatidylethanolamine; LPI: lysophosphatidylinositol.

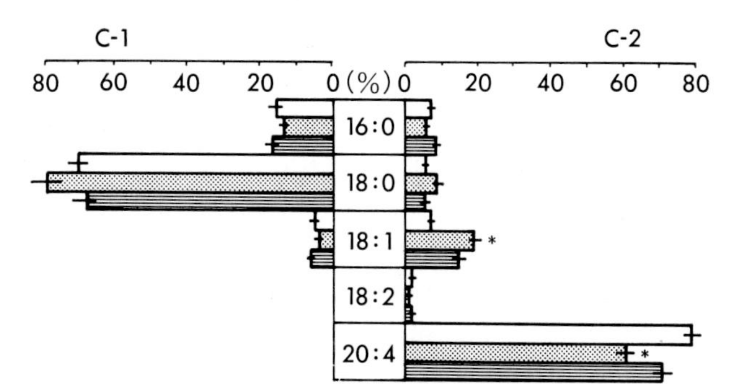

Fig. 2. Changes in the positional distribution of fatty acids in phosphatidylinositol after exposure of human platelets to thrombin. Washed human platelets were exposed to thrombin (2 units/10^9 cells) for the time indicated at 37°C. The positional distribution of fatty acids was determined as described previously (*12*). ▭ non-activated; ▨ activated, 30 sec; ▤ activated, 10 min. *$p<0.01$.

in rabbit neutrophils stimulated with formylmethionyl-leucyl-phenylalanine (*36*), while no lysophospholipid production is detected in mast cells stimulated by various agents to release histamine. This may be due to the high potency to reacylate lysophospholipids by acyltransferase in mast cells.

Thrombin-induced rapid degradation of phospholipids resulting from phospholipase activation was followed by an altered fatty acid pairing solely in PI (Fig. 2), which is ascribed to the involvement of the *de novo* pathway (*12*). However, accompanying PI resynthesis the positional distribution was rearranged to the unique molecular species, 1-stearoyl/2-arachidonoyl, by a deacylation (phospholipase)-reacylation (acyltransferase) system (*12*, *21*). The modification of fatty acid pairing of PI coupled with PI breakdown was also found to occur in formylmethionyl-leucyl-phenylalanine-activated neutrophils (*36*).

II. CALCIUM MOVEMENT

Recently calcium has emerged as a pivotal ion in the regulation of cell function. Briefly, as with all other well studied cell types, calcium movement within a platelet is responsible for initiating events which induce platelet functions (*10*). Level of calcium in the cytosol of cells is low in the resting state, and an elevation of cytosolic calcium triggers biochemical changes which result in cell activation. Increase in cytosolic calcium depends on release of Ca^{2+} from the intracellular calcium-pool and/or the influx of extracellular calcium through the plasma membrane. Feinstein (*9*) examined the time course of intracellular release of membrane-bound Ca^{2+} employing chlorotetracycline fluorescence and found that the stimulant-induced release of intracellular membrane-bound calcium precedes the onset of secretion.

Changes in the ability of platelets to take up Ca^{2+} following stimulation by thrombin have been presented elsewhere (*15*). Exposure of platelets to thrombin in the presence of $^{45}Ca^{2+}$ induced an accumulation of cellular $^{45}Ca^{2+}$, and reached a plateau upon further incubation (Fig. 3). The activity of calcium uptake by activated platelets, measured by $^{45}Ca^{2+}$-pulse labeling, was observed to markedly increase and reach a maximum at 20 sec (*15*). The incorporated $^{45}Ca^{2+}$ was found to be located mainly in the cytosolic fraction (Table I). These findings demon-

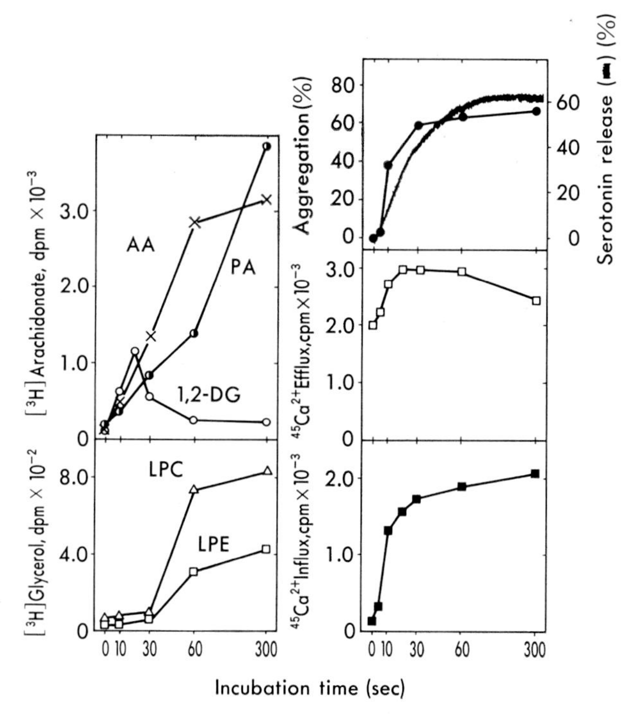

Fig. 3. Time-sequential relationship of platelet responses (serotonin release and aggregation), lipid metabolism and Ca^{2+} flux. All experiments were performed at 24°C. AA: arachidonic acid and its metabolites; PA: phosphatidic acid; 1,2-DG: 1,2-diacylglycerol; LPC: lysophosphatidylcholine; LPE: lysophosphatidylethanolamine.

TABLE I
Distribution of Incorporated $^{45}Ca^{2+}$ in Human Platelet Fraction

Fraction	$^{45}Ca^{2+}$, cpm	(%)
Homogenate	95165	
40,000 × g pellet	8169	(8.9)
100,000 × g pellet	2312	(2.5)
100,000 × g supernatant	80947	(88.2)

After 5 sec incubation of platelets (10^{10} cells) with thrombin (5 units), 30 nCi of $^{45}Ca^{2+}$ was added and the reaction was terminated by adding cold 5 mM EDTA after an additional 5 sec incubation (*15*).

strate that an increase in the cytosolic free Ca^{2+} may in part be due to an enhanced influx of extracellular Ca^{2+}.

The time-sequential relationship between Ca^{2+} flux, phospholipid metabolism and platelet activation is summarized in Fig. 3. The conversion of 1,2-DG to PA and the enhancement of Ca^{2+} uptake are closely associated, which is in good agreement with the hypothesis that PA may act as an endogenous ionophore in various cells (*32*). An increment of cytosolic free Ca^{2+} caused by the enhanced Ca^{2+} influx appeared to precede the formation of lysophospholipids, resulting in the activation of phospholipase A_2 which requires Ca^{2+}, and arachidonic acid liberation.

III. POLYPHOSPHOINOSITIDE METABOLISM

Based on the high affinity of the polyphosphoinositides for Ca^{2+}, it has also been proposed that Ca^{2+} might be released as a direct result of the breakdown of these lipids (*27*). In particular, triphosphoinositide (TPI) binds calcium more strongly than mono- and di-phosphoinositides and its disappearance in membrane might affect various metabolic processes by liberating its bound Ca^{2+} into the cytosol (*9*) and/or perturbing the plasma membrane (*8*). More recently, stimulation-associated rapid changes of TPI have been observed in hepatocytes (*22*), adrenal medulla (*8*) and parotid acinar cells (*35*). In platelets, rapid and transient loss of TPI has also been reported by some groups (*1, 16, 34*). The

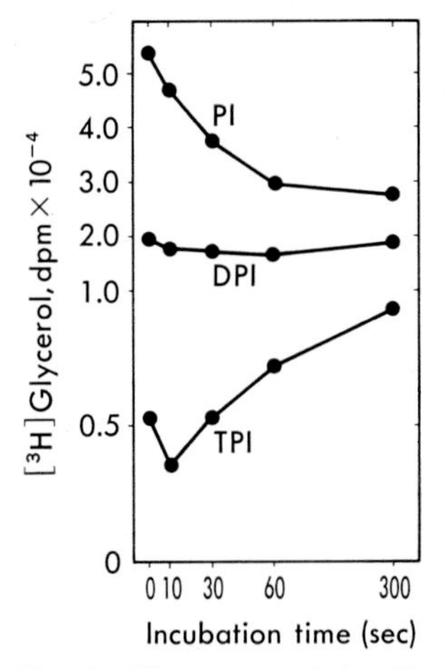

Fig. 4. Time-course of the effects of thrombin on phosphoinositides metabolism in human platelets. [^{3}H]Glycerol-labeled human platelets (10^9 cells) were incubated with thrombin (2 units) at 37°C for the time indicated. PI: phosphatidylinositol; DPI: PI-4-phosphate; TPI: PI-4,5-bisphosphate.

addition of thrombin induces initial loss and subsequent increase of TPI (Fig. 4) without any significant change in diphosphoinositide (DPI) (*16*). A transient loss of TPI was accompanied by concurrent production of 1,2-DG (Fig. 1), suggesting that the initial decrease of TPI might result from its rapid conversion to 1,2-DG *via* TPI phosphodiesterase. There was little incorporation of [^{32}P] phosphate into TPI up to 30 sec, but shortly thereafter a substantial increase of radioactivity was noted. The uptake of [^{32}P]phosphate occurred in parallel with an increase in TPI content (*16*). The initial changes may also be caused by other platelet agonists such as platelet-activating factor (PAF) (*3*) and ADP (*34*), which are known to mobilize intracellular Ca^{2+}. A similar receptor-mediated loss of TPI has been observed after stimulation of hepatocytes (*22*) and parotid acinar cells (*35*), and this phosphodiesteratic cleavage of TPI is independent of Ca^{2+} mobilization. These findings provide evidence that initial loss of TPI is linked to the initiation of cellular calcium elevation by mobilizing plasma membrane-bound Ca^{2+}.

IV. PHOSPHOLIPID BIOSYNTHESIS

All factors which lead to increased levels of cAMP within platelets are known to prevent or reverse thrombin-dependent platelet responses (*e.g.*, serotonin release, phospholipase activation, arachidonic acid liberation and Ca^{2+} influx) (*17*, *24*). However, as shown in Fig. 5, the dibutyryl cAMP-loaded platelets produced a marked enhancement in thrombin-induced incorporation of [^{3}H]glycerol into phosphatidic acid and other phospholipids; there was a 2–4-fold increase compared to thrombin activation alone. Similar results were obtained from platelets pretreated with forskolin, a potent activator of adenylate cyclase, which induced a profound increase in cAMP content of platelets in a dose-dependent fashion.* These results indicate that the critical factor responsible for initiating enhancement of thrombin-induced *de novo* synthesis of glycerolipids from [^{3}H]glycerol appears to be a rise in cAMP level in platelets. Since upon thrombin stimulation the accentuated labeling was observed in all phospholipids in cAMP-loaded platelets, the con-

* Manuscript in preparation by the authors.

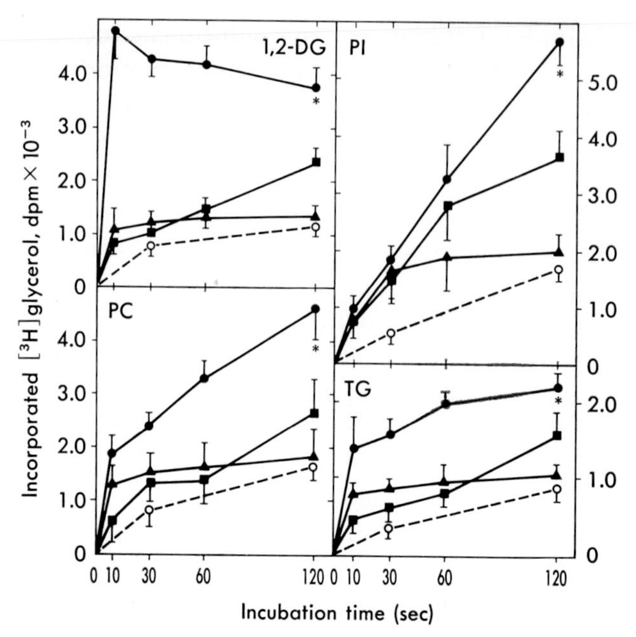

Fig. 5. Effects of dibutyryl cAMP on thrombin-induced [^{3}H]glycerol incorporation into human platelet glycerolipids. Washed platelets (10^8 cells) were preincubated with or without bidutyryl cAMP (2.5 mM) for 5 min at 37°C. They were then exposed to [^{3}H]glycerol (30 μCi) in the presence or absence of thrombin (0.2 units) at 37°C for the indicated interval. *$p<0.001$.

version of glycerol to PA by glycerol kinase seems to be a plausible candidate through which the combination of cAMP and thrombin regulates lipid synthesis. In contract to platelets, a significant enhancement of [^{3}H]glycerol incorporation into PA and PI was observed during histamine release in mast cells (*19*).

SUMMARY

Membrane phospholipid breakdown was observed as an early and crucial response of platelets to chemical stimuli such as collagen and thrombin. Possible mechanisms suggested to be involved in the initial breakdown event are (1) phospholipase A_2 activity to yield lysophospholipids, (2) phospholipase C acting selectively on PI to produce 1,2-DG, and (3) phosphodiesteratic cleavage of TPI to 1,2-DG. The product 1,2-DG was subsequently converted to PA followed by resynthesis to

PI. A marked enhancement in $^{45}Ca^{2+}$ influx occurred in parallel with the conversion of 1,2-DG to PA, and was observed to commence prior to the onset of phospholipase A_2 activation and consequent arachidonate liberation and aggregation. The incorporated $^{45}Ca^{2+}$ was located mainly in the cytosolic fraction. The close parallelism of PI metabolism and Ca^{2+} influx and their precedence over processes of platelet responses indicate that modulation of platelet activation by the increment of cytosolic Ca^{2+} may be mediated at least in part by PA induced-enhanced Ca^{2+} influx.

On the other hand, upon addition of thrombin to platelets with elevated cellular cAMP content as a result of pretreatment with dibutyryl cAMP or forskolin, *de novo* synthesis of phospholipids from [^{3}H]glycerol was enhanced 2–4-fold compared to thrombin stimulation of untreated platelets. In the former case, phospholipid degradation and platelet responses were suppressed. Platelets, prevented from revealing responses by cAMP, instead show acceleration of phospholipid biosynthesis, *i.e.*, phospholipid biosynthesis appears to play a role in suppressing platelet response.

REFERENCES

1 Billah, M.M. and Lapetina, E.G. (1982). *Biochem. Biophys. Res. Commun.* **109**, 217–222.
2 Billah, M.M. and Lapetina, E.G. (1982). *J. Biol. Chem.* **257**, 5196–5200.
3 Billah, M.M. and Lapetina, E.G. (1983). *Proc. Natl. Acad. Sci. U.S.* **80**, 965–968.
4 Bills, T.K., Smith, J.B., and Silver, M.J. (1977). *J. Clin. Invest.* **60**, 1–6.
5 Broekman, M.J., Ward, J.W., and Marcus, A.J. (1980). *J. Clin. Invest.* **66**, 275–283.
6 Cheung, W.Y. (1980). *Science* **207**, 19–27.
7 Cockcroft, S. and Gomperts, B.D. (1979). *Biochem. J.* **178**, 681–687.
8 Farese, R.V., Sabir, A.M., and Vandor, S.L. (1981). *J. Biol. Chem.* **254**, 6842–6844.
9 Feinstein, M.B. (1980). *Biochem. Biophys. Res. Commun.* **93**, 593–600.
10 Gerrard, J.M., Peterson, D.A., and White, J.G. (1982). In *Platelets in Biology and Pathology*, vol. 2, ed. Gordon, J.L., pp. 349–372. Amsterdam: Elsevier.
11 Hokin-Neaverson, M.R. (1974). *Biochem. Biophys. Res. Commun.* **58**, 763–768.
12 Imai, A., Yano, K., Kameyama, Y., and Nozawa, Y. (1981). *Biochem. Biophys. Res. Commun.* **103**, 1092–1099.
13 Imai, A., Yano, K., Kameyama, Y., and Nozawa, Y. (1982). *Japan. J. Exp. Med.* **52**, 99–105.
14 Imai, A. and Nozawa, Y. (1982). *Biochem. Biophys. Res. Commun.* **105**, 693–700.
15 Imai, A., Ishizuka, Y., Kawai, K., and Nozawa, Y. (1982). *Biochem. Biophys. Res. Commun.* **108**, 752–759.
16 Imai, A., Nakashima, S., and Nozawa, Y. (1983). *Biochem. Biophys. Res. Commun.* **110**, 108–115.

17 Imai, A., Hattori, H., Takahashi, M., and Nozawa, Y. (1983). *Biochem. Biophys. Res. Commun.* **112**, 693–700.
18 Imai, A., Takahashi, M., and Nozawa, Y. (1983). *Cryobiology*, **21**, 255–259.
19 Ishizuka, Y., Imai, A., Nakashima, S., and Nozawa, Y. (1983). *Biochem. Biophys. Res. Commun.* **111**, 581–587.
20 Jones, L.M. and Michell, R.H. (1974). *Biochem. J.* **142**, 583–590.
21 Kameyama, Y., Yoshioka, S., Imai, A., and Nozawa, Y. (1983). *Biochim. Biophys. Acta* **752**, 244–250.
22 Kirk, S.J., Creba, J.A., Downes, C.P., and Michell, R.H. (1981). *Biochem. Soc. Trans.* **9**, 377–379.
23 Kohn, L.D. (1982). *Hormone and Receptors*, Chichester: John Wiley and Sons.
24 Lapetina, E.G., Schmitges, C.J., Chandrabose, K., and Cuatrecasas, P. (1977). *Biochem. Biophys. Res. Commun.* **76**, 828–835.
25 Lunt, G.G. and Pickard, M.R. (1975). *J. Neurochem.* **24**, 1203–1208.
26 McKean, M.L., Smith, J.B., and Silver, M.J. (1981). *J. Biol. Chem.* **256**, 1522–1524.
27 Michell, R.H. (1975). *Biochim. Biophys. Acta* **415**, 81–147.
28 Ohki, K., Yamauchi, T., Banno, Y., and Nozawa, Y. (1981). *Biochem. Biophys. Res. Commun.* **100**, 321–327.
29 Oron, Y., Lowe, M., and Selinger, Z. (1975). *Mol. Pharmacol.* **11**, 79–86.
30 Rittenhouse-Simmons, S. (1979). *J. Clin. Invest.* **63**, 580–587.
31 Rittenhouse-Simmons, S. and Deykin, D. (1981). In *Platelets in Biology and Pathology*, vol. 2, ed. Gordon, J.L., pp. 349–372. Amsterdam: Elsevier.
32 Serhan, C., Anderson, P., Goodman, E., Dunham, P., and Weissman, G. (1982). *J. Biol. Chem.* **257**, 2736–2741.
33 Takai, Y., Kishimoto, A., Kawahara, Y., Minakuchi, R., Sano, K., Kikkawa, U., Mori, T., Yu, B., Kaibuchi, K., and Nishizuka, Y. (1981). *Adv. Cyclic Nucleotide Res.* **14**, 301–313.
34 Vickers, J.D., Kinlough-Rathbone, P.L., and Mustard, J.F. (1982). *Blood* **60**, 1247–1250.
35 Weiss, S.J., McKinney, J.S., and Putney, J.W., Jr. (1982). *Biochem. J.* **206**, 555–560.
36 Yano, K., Hattori, H., Imai, A., and Nozawa, Y. (1983). *Biochim. Biophys. Acta* **752**, 137–144.

2

PROPERTIES AND FUNCTIONS OF ENZYMES HYDROLYSING INOSITOL PHOSPHOLIPID

R. F. IRVINE, A. J. LETCHER, D. J. LANDER,
AND R. M. C. DAWSON

Department of Biochemistry, ARC Institute of Animal Physiology, Babraham, Cambridge CB2 4AT, U.K.

The stimulated metabolism of the phosphoinositides Ptd Ins, Ptd Ins 4P and Ptd Ins 4,5P_2* is a widespread phenomenon in agonist-activated tissues (*6*, *13*, *21*, *22*). Its various cellular functions (of which there may be more than one (*13*)) are not definitely established as yet, but there is a remarkable coincidence between calcium-mobilizing agonists and this response (*21*), and recently some direct evidence has been provided that Ins 1,4,5P_3, one of the products of inositide catabolism, can mobilize calcium from intracellular stores (*25*). Another product of the catabolism of inositides, diacylglycerol, can act as a stimulant to a protein kinase (*17*) or as a source of arachidonic acid (*3*, *11*, *20*) and the interactions between the possible calcium-releasing and diacylglycerol-forming aspects of inositide catabolism is an area under investigation in several laboratories (*8*, *17*).

It is generally accepted that the initial event in the stimulated

* Abbreviations: Ptd Ins, Phosphatidylinositol; Ptd Ins 4P, Phosphatidylinositol 4′ phosphate; Ptd Ins 4,5P_2, phosphatidylinositol 4′,5′ bis phosphate; Ptd Cho, Phosphatidylcholine; Ptd Ethan, Phosphatidylethanolamine; Ins 1P, Inositol 1 phosphate, Ins 1,4P_2, Inositol 1,4 bis phosphate; Ins 1,4,5P_3, Inositol 1,4,5 tris phosphate.

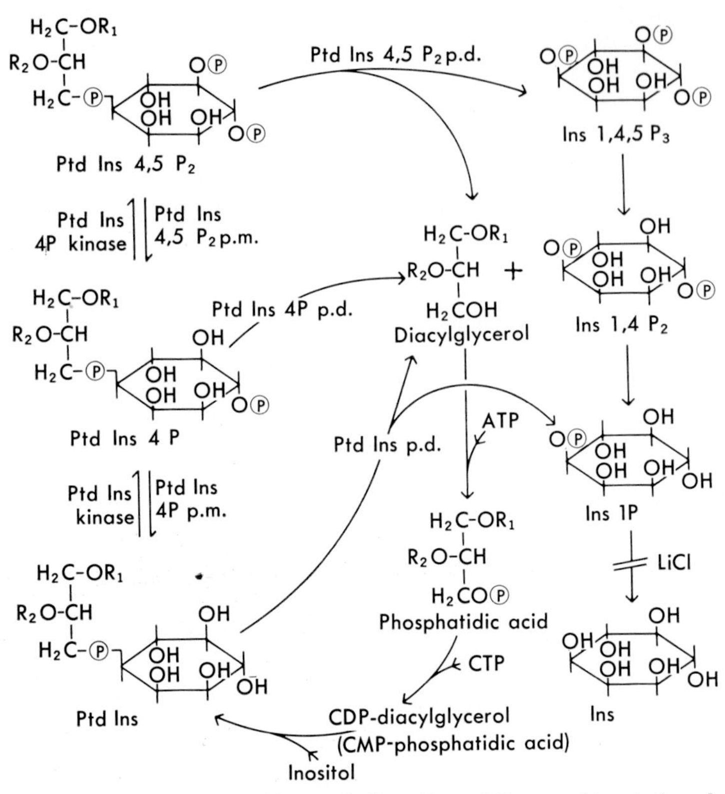

Fig. 1. Pathways of inositide metabolism. R_1 and R_2 are abbreviations for fatty acyl chains. R_1 is predominantly stearoyl and R_2 arachidonoyl. Ⓟ is an abbreviation for phosphate group. P.d.; phosphodiesterase, P.m.; phosphomonoesterase.

metabolism of inositides is a phosphodiesterasic cleavage of an inositol lipid (*21*). Until recently the substrate for this cleavage was regarded as solely Ptd Ins, but more recent evidence (*1*, *4*, *5*) suggests that the initial hydrolysis may be of polyphosphoinositides, Ptd Ins 4P and Ptd Ins $4{,}5P_2$ (Fig. 1). Whether Ptd Ins phosphodiesterase is also activated is as yet uncertain. It is perhaps relevant to add that under well-defined conditions Ptd Ins phosphodiesterase and polyphosphoinositide phosphodiesterase activities have not yet been separated (*2*, *18*, *24*, but see *26*), and recently Low and Weglicki (*19*) have provided some strong circumstantial evidence that the same proteins (the enzyme exists in several forms (*9*, *10*, *19*)) catalyse the hydrolysis of all inositol lipids. Thus, when we study the Ptd Ins phosphodiesterase and Ptd Ins $4{,}5P_2$ phosphodiesterase activities of a crude rat brain supernatant, we may be looking at different activities of the same proteins.

I. PTD INS PHOSPHODIESTERASE

We have reviewed the work done by this laboratory on this enzyme elsewhere (*12, 13*), and only a brief summary is necessary here. When assayed at 1 mM Ca^{2+}, Ptd Ins phosphodiesterase has the potential to degrade all its available substrate in a very short time, and is prevented from doing so by the presence of positively charged proteins, KCl, and, probably most important, choline-containing phospholipids in the membrane. The latter effect (inhibition by Ptd Cho) may explain why several laboratories have reported a membrane-bound Ptd Ins $4,5P_2$ phosphodiesterase activity (assayed at 1 mM Ca^{2+}) but no Ptd Ins phosphodiesterase (see ref. *12*); the former enzyme can readily attack its substrate in a biological membrane with 1 mM Ca^{2+} present (*7, 14, 15*).

At 1 μM Ca^{2+}, the optimum pH for hydrolysis of Ptd Ins shifts from about 5 to nearer 7 (*15, 16*) unless sufficient divalent cations are present to neutralize the charge on the substrate (Fig. 2). Ptd Ins $4,5P_2$ phosphodiesterase does not share this phenomenon (*14*), and

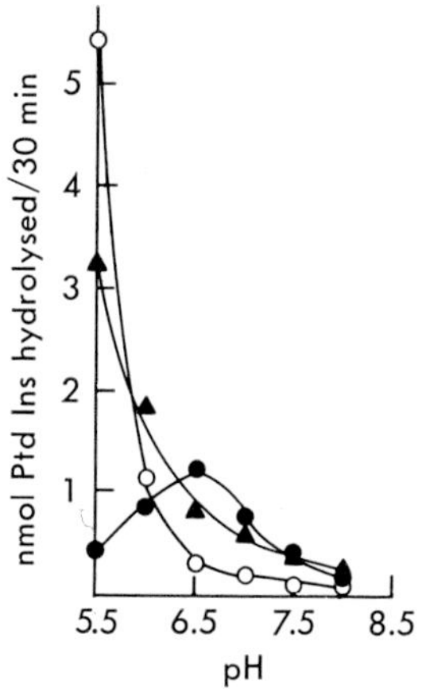

Fig. 2. Effect of ionic environment on pH optimum of Ptd Ins phosphodiesterase. A fraction from a chromatofocusing column (*9, 16*) containing only activity at pH 5.5, 1 mM Ca^{2+} was assayed at: 1 mM Ca^{2+} (○); 1 μM Ca^{2+} (Ca^{2+}/EDTA buffer) (●); 1 μM Ca^{2+}, 1 mM Mg^{2+}, 80 mM KCl (Ca^{2+}/EGTA buffer) (▲).

under various ionic environments consistently shows two pH optima at about 5 and 8, as does Ptd Ins phosphodiesterase at 1 mM Ca^{2+}. There are factors in the brain supernatant which inhibit the Ptd Ins phosphodiesterase under limiting calcium concentrations, and which can be partially removed by trypsin treatment (*15*, *16*). If Mg^{2+} (1 mM) and KCl (80 mM) are present in the assay (*i.e.*, approximately the ionic conditions found *in vivo*), then addition of Ca^{2+} from 10^{-8} to 10^{-4} M makes little difference to Ptd Ins phosphodiesterase activity (*15*). In summary, Ptd Ins phosphodiesterase is unlikely to be limited by calcium *in vivo*, but is instead probably controlled by the state of presentation of its substrate.

II. PTD INS 4,5P$_2$ PHOSPHODIESTERASE AND PHOSPHOMONOESTERASES

The basic properties of these enzymes have been reviewed elsewhere (*12*), and our recent work on the enzyme, published in ref. *14* concentrates on possible control mechanisms which might operate *in vivo*. The phosphorylation and dephosphorylation of Ptd Ins (Fig. 1) and its derivatives are cellular events which are occurring very rapidly indeed; recent measurements of the equilibration of the y-phosphate of ATP with the 4′ and 5′ phosphates of Ptd Ins 4,5P$_2$ in hepatocytes show that the rate of equilibration is too fast to be measured by present techniques (*i.e.*, of the order of a minute or two, P. T. Hawkins personal communication). Thus, if a mechanism could be found for the switching on of the Ptd Ins 4,5P$_2$ phosphodiesterase, this would not only explain the rapid decline in Ptd Ins 4,5P$_2$ in stimulated cells, but would quite probably be sufficient to account also for the disappearance of Ptd Ins (by phosphorylation, ref. *23*).

We have found that although the Ptd Ins 4,5P$_2$ phosphodiesterase will hydrolyse its substrate when the latter is part of a membrane at 1 μM Ca^{2+} with no KCl or Mg^{2+} present, it will not do so until the Ca^{2+} is raised to millimolar levels if Mg^{2+} and KCl are present at concentrations similar to those pertaining *in vivo* (*14*). In this respect it resembles closely the membrane-bound Ptd Ins 4,5P$_2$ phosphodiesterase activity found in erythrocyte ghosts (*7*). As mentioned above, this ability to attack a membrane with 1 mM Ca^{2+} present contrasts with Ptd

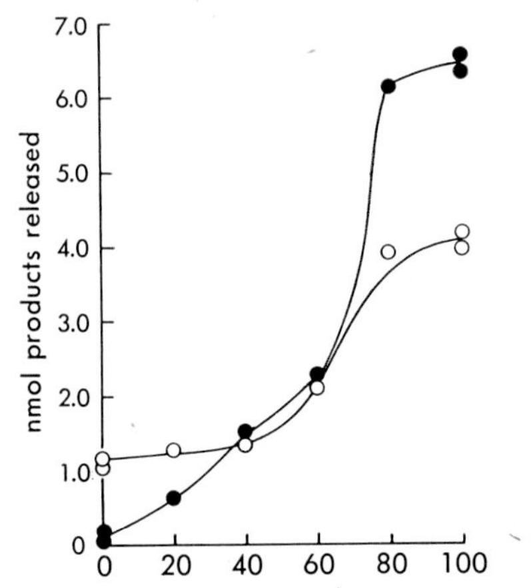

Fig. 3. Effect of substrate configuration on Ptd Ins 4,$5P_2$ phosphodiesterase and phosphomonoesterase activity. Enzyme activities were assayed in a Ca^{2+}/EGTA buffer containing 1 μM Ca^{2+}, 1 mM Mg^{2+} and 80 mM KCl at pH 5.5. The substrate (^{32}P-labelled Ptd Ins 4,$5P_2$) was mixed with a ten-fold excess of a Ptd Cho/Ptd Ethan mixture of varying composition, and was assayed for Ptd Ins 4,$5P_2$ phosphomonoesterase and diesterase activities as in ref. *14*. ○; phosphomonoesterase, ●; phosphodiesterase.

Ins phosphodiesterase activity which is inhibited by Ptd Cho at all Ca^{2+} concentrations (*12*).

If however, the conformation of the substrate is changed from a bilayer to a non-bilayer configuration (by admixing with Ptd Ethan), then a dramatic lowering occurs in the Ca^{2+} dependency, resulting in a stimulation of the phosphodiesterase at 1μM Ca^{2+} (+1 mM Mg^{2+} and 80 mM KCl) so that it is similar in activity to the monoesterase (*14* and Fig. 3). We have suggested (*14*) that a change in membrane structure of a similar sort takes place in the location of the agonist receptor, and that this alone is sufficient to account for the stimulated phosphodiesteric cleavage of Ptd Ins 4,$5P_2$ which occurs on cell stimulation (see Introduction).

It is relevant to add that if presenting the substrate in a non-bilayer configuration at 1 μM Ca^{2+}, 1 mM Mg^{2+} and 80 mM KCl is indeed similar to the "activated" substrate *in vivo*, then under these conditions the enzymes in rat brain at pH 5.5 and 7.25 show a hundred-fold preference for Ptd Ins 4,$5P_2$ over Ptd Ins (*14*). Thus, a receptor-mediated change in lipid presentation could alone generate diacylglycerol, with its protein-kinase stimulating properties (*17*) and Ins $1,4,5P_3$ with its calcium-releasing properties (*25*).

SUMMARY

The properties of Ptd Ins $4,5P_2$ phosphodiesterase and Ptd Ins phosphodiesterase activities in rat brain have been studied. The ionic environment of the assay can have profound effects on the calcium-dependency and pH requirements of the enzyme activities. Under an approximation of physiological conditions (1 μM Ca^{2+}, 1 mM Mg^{2+} and 80 mM KCl), Ptd Ins $4,5P_2$ phosphodiesterase cannot hydrolyse its substrate unless the latter is presented as a non bilayer configuration. Under such conditions (and given such non-bilayer substrate) a rat brain supernatant shows a pronounced specificity for Ptd Ins $4,5P_2$ over Ptd Ins.

REFERENCES

1 Agranoff, B.W., Murthy, P., and Seguin, F.B. (1983). *J. Biol. Chem.* **258**, 2076–2078.
2 Atherton, R.S. and Hawthorne, J.N. (1968). *Eur. J. Biochem.* **4**, 68–75.
3 Bell, R.L., Kennedy, D.A., Standford, N., and Majerus, P.W. (1979). *Proc. Natl. Acad. Sci. U.S.* **76**, 3238–3241.
4 Berridge, M.J. (1983). *Biochem. J.* **212**, 849–858.
5 Berridge, M.J., Dawson, R.N.C., Downes, C.P., Heslop, J.P., and Irvine, R.F. (1983). *Biochem. J.* **212**, 473–482.
6 Downes, C.P. and Michell, R.H. (1982). *Cell Calcium* **3**, 467–502.
7 Downes, C.P. and Michell, R.H. (1983). *Biochem. J.* **302**, 53–58.
8 Hallam, T., Sanchez, A., and Rink, T.J. (1983). *Nature* **305**, 317–319.
9 Hirasawa, K., Irvine, R.F., and Dawson, R.M.C. (1982). *Biochem. J.* **205**, 437–442.
10 Hoffman, S.L. and Majerus, P.W. (1982). *J. Biol. Chem.* **257**, 6461–6469.
11 Irvine, R.F. (1982). *Biochem. J.* **204**, 3–16.
12 Irvine, R.F. (1982). *Cell Calcium* **3**, 295–309.
13 Irvine, R.F., Dawson, R.M.C., and Freinkel, N. (1982). In *Contemporary Metabolism*, vol. 2, ed. Freinkel, N., pp. 301–342. New York: Plenum Press.
14 Irvine, R.F., Letcher, A.J., and Dawson, R.M.C. (1983). *Biochem. J.* in press.
15 Irvine, R.F., Letcher, A.J., and Dawson, R.N.C. (1984). In *Physiological Role of Phospholipids in the Nervous System*, ed. Horrocks, L.A., Kanfer, J.N., and Porcellati, G.A., New York: Raven Press, in press.
16 Irvine, R.F. and Dawson, R.M.C. (1983). *Biochem. J.* **215**, 431–432.
17 Kaibuchi, K., Sano, K., Hoshijima, M., Takai, Y., and Nishizuka, Y. (1982). *Cell Calcium* **3**, 323–335.
18 Kemp, P., Hubscher, G., and Hawthorne, J.N. (1961). *Biochem. J.* **79**, 193–200.

19 Low, M.G. and Weglicki, W.B. (1983). *Biochem. J.* **215**, 325–334.
20 Mauco, G., Chap, H., Simon, M.F., and Douste-Blazy, C. (1978). *Biochemie* **60**, 653–661.
21 Michell, R.H. (1975). *Biochim. Biophys. Acta* **415**, 81–147.
22 Michell, R.H. (1982). *Cell Calcium*, **3**, 285–294.
23 Michell, R.H., Kirk, C.J., Jones, L.M., Downes, C.P., and Creba, J.A. (1981). *Proc. Roy. Soc. Lond. B.* **296**, 123–137.
24 Rittenhouse, S. (1983). *Proc. Natl. Acad. Sci. U.S.* **80**, 5417–5420.
25 Streb, H., Irvine, R.F., Berridge, M.J., and Schulz, I. (1983). *Nature* **306**, 67–69.
26 Thompson, W. and Dawson, R.M.C. (1964). *Biochem. J.* **91**, 237–243.

3

POSSIBLE INVOLVEMENT OF PROTEASES IN PHOSPHATIDYLINOSITOL TURNOVER*

KEISUKE HIRASAWA

Department of Cell Biology, National Institute for Basic Biology, Okazaki 444, Japan

One effect of acetylcholine on pancreas slices was to cause changes in phospholipid metabolism, a rapid, specific and large incorporation of ^{32}P into phosphatidylinositol and phosphatidic acid (*27*). Freinkel (*18*) demonstrated a similar increase in the turnover of phosphatidylinositol in thyroid after the addition of thyrotropin. A specific increase in the turnover of phosphatidylinositol has since been documented in a number of tissues in response to stimulation by agonists, *i.e.*, acetylcholine, neurotransmitter, hormones and other agents.

It has been generally accepted that stimulation of cells is initiated by the interaction of receptor and agonist, which can sometimes induce an increase in the catabolism of the inositol phospholipids, in turn bringing about an increased synthesis. Stimulated phosphatidylinositol turnover is a crucial step in the complex series of coordinated processes which increase the physiological responses of a cell. It may represent a universal multifunctional transducing mechanism (*6*).

* This work was done at the Department of Biochemistry, ARC, Institute of Animal Physiology, Cambridge, U.K. All correspondence should be addressed to the Department of Biochemistry, Meiji Institute of Health Science, Naruda, Odawara 250, Japan.

The activation of the receptors involved, for example, muscarinic rather than nicotinic and α-rather than β-adrenergic, which constitute a class distinct from those acting through cAMP, results in the increased availability of the other second messengers. In no system does cAMP or its derivatives elicit the phospholipid metabolic effects of stimuli. The precise consequences and physiological significance of the phospholipid effects have heretofore been unknown. Many functions of the increased phosphatidylinositol turnover, however, have been proposed: membrane fusion (*40*), early events in cell division (*17*), protein kinase activation (*39, 46, 47*), calcium gating (*37*), arachidonic acid release (*5*) and increases in cGMP levels (*47*). It seems likely that the putative initial stage of increased phosphatidylinositol turnover on stimulation is by hydrolysis of phosphatidylinositol by its specific phospholipase C.

The basic properties of the enzyme, substrate specificity, the hydrolysis products and calcium-dependency have been well demonstrated in various mammalian tissues, sheep pancreas (*8, 13*), rat liver (*31*), guinea pig mucosa (*4*), rat brain (*22*) and pig lymphocytes (*2, 3*). The phospholipase C is calcium-ion dependent, but it is not known whether changes in Ca^{2+} concentrations in the physiological range could exert a controlling influence.

Dawson and colleagues (*9, 10*) have identified the cleavage products of the enzyme as diglyceride, D-inositol-1-phosphate and D-inositol-1,2-cyclicphosphate; the latter can be further hydrolysed to D-inositol-1-phosphate. As yet no specific physiological function for the water-soluble products has been uncovered. Nishizuka and his colleagues (*39, 46, 47*) have well demonstrated that diacylglycerol as novel intracellular messenger stimulated the phospholipid-Ca^{2+}-dependent protein kinase *in vivo* and *in vitro*.

I. INHIBITORY EFFECT OF PHOSPHATIDYLCHOLINE ON Ca^{2+}-DEPENDENT PHOSPHATIDYLINOSITOL PHOSPHOLIPASE C

Low and Finean (*34*) and Irvine and Dawson (*29*) have observed that phosphatidylinositol phospholipase C will not readily attack its substrate when this is a constituent of an isolated cell membrane. They reported some possible inhibitory factors of the enzyme acting *in vivo* as well as *in vitro*. These inhibitory factors have been demonstrated to be KCl and $MgCl_2$ at a physiological concentration (*2*) and, probably the most important, choline-containing phospholipids (phosphatidylcholine, sphingomyelin and lysophosphatidylcholine) (*11*).

In contrast, acidic phospholipids such as phosphatidic acid can act as potent activators of phosphatidylinositol phospholipase C (*11*, *28*). These effects of other phospholipids have been examined in a monolayer model membrane system with the enzyme acting at an air-water interface (*21*). The inhibitory effects of choline-containing phospholipids can be partially reversed by the addition of sufficient phospholipid activators to the system. Irvine *et al.* (*28*) have suggested that once the cytoplasmic phosphatidylinositol phosphodiesterase has catalysed the breakdown, the reaction could be self-amplified by diglyceride kinase forming phosphatidic acid which is a most potent activator of the enzyme in the test tube, or alternatively by diglyceride lipase forming activating unsaturated fatty acids (principally arachidonic acid).

However, these observations have been obtained under high calcium concentration [mM] because the enzyme was fully activated (Fig. 1). Therefore, there are two negative factors for the Ca^{2+}-dependent phosphatidylinositol phospholipase C activity *in vivo*; one is that relatively high content choline-containing phospholipids are present in the cell membrane, and the other is that intracellular free calcium ion is much too low (10^{-7}–10^{-6} M). Thus an activation factor may exist to

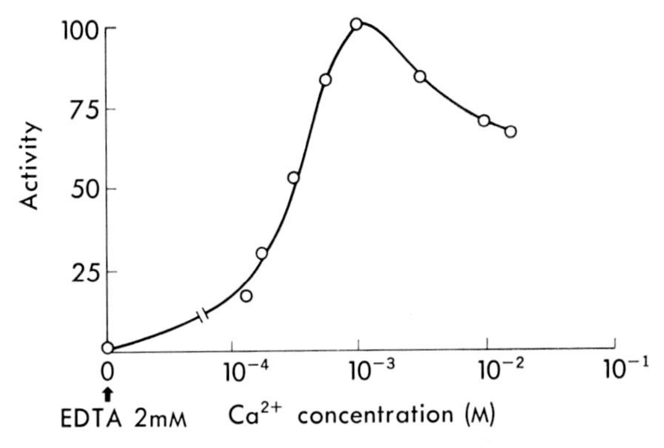

Fig. 1. Effect of Ca^{2+} ion on the hydrolysis of [^{32}P]-phosphatidylinositol monomolecular film at 25×10^{-3} N/m. The activity was expressed as a percentage of the maximum hydrolysis at 1 mM Ca^{2+}.

take away or overcome the various inhibitory effects on the enzyme activity and to alter its requirement for calcium ions.

II. HETEROGENEITY OF Ca^{2+}-DEPENDENT PHOSPHATIDYL-INOSITOL PHOSPHOLIPASE C

Some evidence for at least two distinct pH optima of phosphatidylinositol phospholipase C has been presented in lymphocytes (*2*, *3*) and rat brain (*22*) if the enzyme is assayed at 0.4 mM and 1 mM Ca^{2+} respectively, but only one peak is observed if the calcium concentration is buffered to 1 μM. Kemp *et al.* (*31*) have also reported evidence of two pH optima in the liver enzyme, a major one at 5.7 and a minor one at 6.9. On the other hand, only one peak of activity was observed even with calcium at 1–2 mM, at pH 5.5–5.8 in rat brain (*50*), guinea pig intestinal mucosa (*4*), ox brain (*32*) and rabbit smooth muscle (*1*), and at pH 7 for the enzyme purified from rat liver (*46*). Dawson *et al.* (*13*) have recently extended these observations to a sheep pancreas supernatant fraction where strong phosphatidylinositol phospholipase C activity is seen up to pH 8.5. The similar situation exists in rat liver and kidney (*24*). In addition, the heterogeneity of the phospholipase C has been examined by standard isoelectric focusing and by a new chromatofocusing technique with an ampholine pH gradient buffer. Chromatofocusing has proved to be a valuable aid in probing the heterogeneity of the present enzyme. Highly reproducible patterns of enzyme elution were obtained from each tissue. That the revealed fractions are not artefacts of the method is indicated by the broadly similar patterns obtained for chromatofocusing and electrofocusing. When the fractions of the chromatofocusing were assayed at pH 5.5 with 1 mM Ca^{2+}, four peaks were eluted: minor, but consistent, activity at the beginning of the elution with a pI of near 7.2 or above; a second peak at pH 6.0–5.85; a third broad peak with a wide range pH 5.3–4.2; and a fourth peak which was eluted with 1 M NaCl, suggesting an isoenzyme with a pI 3.8. (supported by the result of isoelectric focusing).

If all the chromatofocusing fractions were assayed at pH 7.0 or 8.0 (and 1 mM Ca^{2+}), only a single sharp peak was detected, with a pI of 4.6–4.8. Takenawa and Nagai have presumably isolated this isoenzyme fraction from rat liver supernatant (*48*). Similarly, two distinct phosphatidylinositol phospholipase C were purified from sheep seminal vesicular glands by Hofmann and Majerus (*26*). According to the isoelectric point of the two isomers, these are included in the chromatofocusing result of rat brain (*23*), liver and kidney (*24*). Hirasawa *et al.* (*23*) have shown that the phosphatidylinositol phospholipase C has at least four heterogeneous isoenzymes.

III. TRYPSIN-TREATMENT OF Ca^{2+}-DEPENDENT PHOSPHATIDYLINOSITOL PHOSPHOLIPASE C

Normally the enzyme extracted from brain is virtually inactive under physiological calcium concentrations (10^{-7}–10^{-6} M) at neutral pH. When the control cytosolic supernatant was assayed under 1 μM Ca^{2+} free concentration a small activity with an optimum 6.0 was detected which was fully activated by an increase of calcium concentration up to 1 mM (*22*), a concentration which is much higher than the intracellular level (10^{-7}–10^{-6} M). Furthermore, if phosphatidylcholine is present in a lipid bilayer or monolayer the phosphatidylinositol phospholipase C cannot attack its substrate, even at high calcium concentrations (10^{-3} M). These previous observations suggest that the enzymic hydrolysis of phosphatidylinositol in cell membranes requires activation factors which enable the enzyme to penetrate into the membrane and also to hydrolyse the substrate at low calcium concentrations.

One possible answer to the question of how these events can occur could be suggested by the results that the phosphatidylinositol breakdown under physiological calcium concentrations (10^{-7}–10^{-6} M at pH 7.25) was stimulated by trypsin. This activation is thought to be caused by proteolytic cleavage of the enzyme as demonstrated by the fact that heat denaturation of the trypsin or addition of soya bean trypsin inhibitor prevented the effect (*20*). Presumably the effect of trypsin is to convert the calcium-dependent phosphatidylinositol phospholipase C which is active under 1 mM Ca^{2+} to a new form which is active at low calcium concentrations. It is possible that trypsin potentiates the activity of the phosphatidylinositol phospholipase C by releasing a peptide from the Ca^{2+} insensitive form of the enzyme in a fashion similar to the pancreatic phospholipase A_2 released which is cleaved by trypsin (*14*).

If a brain supernatant was preincubated at 37°C for long periods, the enzyme activity was also stimulated if assayed with 1 μM Ca^{2+} present. This suggested that endogenous proteases possibly activate the enzyme. Such enzymes have been demonstrated in the brain cytosolic fraction (*33*). When both the trypsin-treated and the preincubated cytosolic supernatant from rat brain were applied to a chromatofocusing column (*25*), trypsin-treatment of the cytosolic supernatant changed the heterogeneous chromatofocusing profile of the enzyme activity to a new pattern, with the disappearance of the peaks at pI 5.0, 4.2 and <4 and the appearance of a new peak at pI 6.1. This latter peak was active at pH 7.25 and 1 μM Ca^{2+}. After a one hour preincubation of the supernatant, the heterogeneous profile of the chromatofocusing chromatogram did not change noticeably, but a new peak at pI 5.0 appeared when the activity was assayed at pH 7.25, 1 μM Ca^{2+}. The two peaks fully activated by a free Ca^{2+} in the range 10^{-7}–10^{-6} M showed a pH optimum of about 7.25, and both produced inositol-1,2-cyclic phosphate and inositol-1-phosphate as water-soluble products.

IV. Ca^{2+}-DEPENDENT PHOSPHATIDYLINOSITOL PHOSPHOLIPASE C IN LYMPHOCYTES

A specific increase in phosphatidylinositol turnover and breakdown has been demonstrated in the lymphocytes in response to phytohaemaglutinin (*17*) and concanavalin A (*19*, *35*) accompanied by accumulation of phosphatidic acid and diglyceride.

It has been reported that the Ca^{2+}-dependent soluble phosphatidylinositol phospholipase C obtained from pig lymphocytes was active in the presence of 1 μM Ca^{2+} at pH 7.0. Figure 2 shows the Ca^{2+}-sensitivity of phosphatidylinositol phospholipase C in lymphocytes supernatant. Such characteristics are similar to the activity of certain peaks obtained by chromatofocusing the trypsinized or preincubated cytosolic supernatant from rat brain (*25*). The enzyme forms in lymphocytes separated when the chromatofocusing technique is used (*20*). Chromatofocusing of lymphocyte supernatant showed two different Ca^{2+}-responsive forms of the enzyme appearing at almost identical isoelectric points to the Ca^{2+}-responsive forms produced in trypsin treated and preincubated brain supernatants (*25*). At present, it is unknown whether in

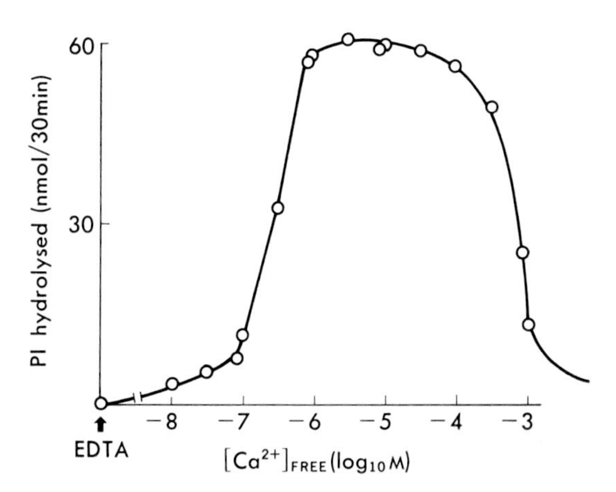

Fig. 2. Effect of free calcium ions on the phosphatidylinositol phosphodiesterase activity in the lymphocytes supernatant. The enzyme activity was assayed at pH 7.25 controlling the free calcium ions by Ca^{2+}-EDTA buffers (Raaflaub, 1956).

lymphocytes the enzyme is present as a form which is active at low calcium concentrations, or whether it is converted to this active form by proteolytic enzymes during preparation and homogenization of the lymphocytes. However, this observation still clearly showed that two active and inactive forms of phosphatidylinositol phospholipase C under physiological calcium concentrations can be obtained in soluble fraction without artificial treatment. In contrast, in the brain preparations phospholipase C only becomes sensitive to low levels of calcium if it is preincubated or treated with trypsin (*25*).

In recent years it has been reported that limited proteolysis by trypsin activates not only phospholipases secreted from the pancreas but also some calmodulin-dependent enzymes such as Ca^{2+}-Mg^{2+}-ATPase (*44*, *48*), cyclic nucleotide phosphodiesterase (*7*), phosphorylase kinase (*15*) and NAD^{+} kinase (*36*).

On the other hand, Seals and Czech (*43*, *44*) have suggested that the insulin second messenger is probably an activation of membrane-bound protease. When rat adipose was stimulated by insulin, 9000 M.W. polypeptide as intracellular mediator was released from the plasma membrane (*43*). It has also been reported that insulin stimulates phosphatidylinositol turnover (*16*). Furthermore, many protease inhibitors block the release of histamine from mast cell (*30*) and of

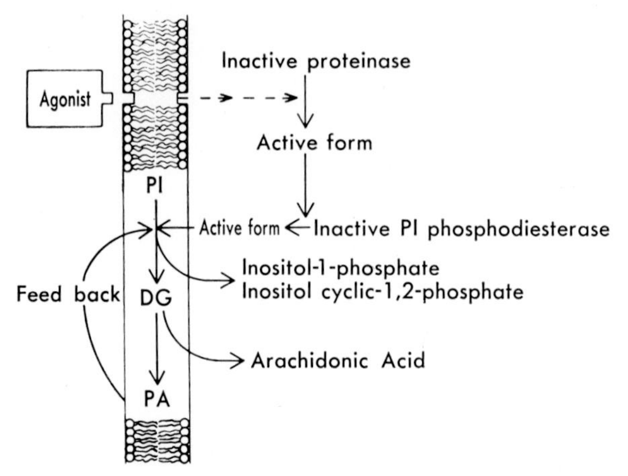

Fig. 3. Proposed scheme of control of phosphatidylinositol hydrolysis.

catecholamine from adrenal medullary cell (*38*). These observations suggest that one early step in phosphatidylinositol breakdown caused by an agonist may be an activation of phosphatidylinositol phospholipase C by a protease (Fig. 3). The possibility that there may be receptor-sensitive proteases is one that needs serious consideration in light of the results.

SUMMARY

Since the original discovery of the Ca^{2+}-dependent phosphatidylinositol phospholipase C in sheep pancreas and rat liver, the enzyme in various mammalian tissues has been assayed at high calcium concentration [mM] for the optimum condition. However, the presence of phosphatidylcholine reduced the ability of the enzyme to penetrate into monolayers of the substrate and catalyse its hydrolysis even with high calcium ion concentrations. Therefore, the enzyme appears not to attack its substrate in a membrane until stimulation occurs.

One can speculate on the control and induction factor requirements of the enzymic hydrolysis of phosphatidylinositol in cell bilayer membrane. In other words, why is it that in a non-stimulated state, little phosphatidylinositol will be hydrolysed, whereas on stimulation the enzyme will attack the substrate in the membrane?

Of significance in the consideration of control mechanisms is that the enzyme in brain when modified by limited proteolysis does not require high calcium concentration for the full activation. If the enzyme acts *in vivo* only at low calcium concentrations probably the most effective activation factor would be trypsin-like cleavage of the enzyme molecule to convert it into this responsive form. The activation of the enzyme by trypsin is not an artefact. Evidence has been obtained of the presence of phosphatidylinositol phospholipase C in such a zymogen form, at least in the brain. The enzyme could be switched on by an endogenous trypsin-like protease which can be demonstrated to operate by preincubation of the brain cytosolic supernatant. However, results obtained by chromatofocusing of lymphocyte supernatant without trypsin treatment showed two different Ca^{2+}-responsive forms of the enzyme appearing at almost identical isoelectric points to the Ca^{2+}-responsive forms produced in brain supernatants. From the evidence described here it seems possible that an early event in phospholipase C type cleavage of phosphatidylinositol is this type of an activation of the enzyme by endogenous proteases. Moreover, the proteolytic activity could possibly be controlled by receptor-stimulation.

Acknowledgments
Prof. Y. Nishizuka is thanked for his criticism and encouragement.

REFERENCES

1 Abdel-Latif, A.A., Luke, B., and Smith, J.P. (1980). *Biochim. Biophys. Acta* **614**, 425–434.
2 Allan, D. and Michell, R.H. (1974). *Biochem. J.* **142**, 591–597.
3 Allan, D. and Michell, R.H. (1974). *Biochem. J.* **142**, 599–604.
4 Atherton, R.S. and Hawthorne, J.N. (1968). *Eur. J. Biochem.* **4**, 68–75.
5 Bell, R.L., Kennerly, D.A., Stanford, N., and Majerus, P.W. (1979). *Proc. Natl. Acad. Sci. U.S.* **76**, 3228–3241.
6 Berridge, M.J. (1981). *Mol. Cell. Endocr.* **24**, 115–140.
7 Cheung, W.Y. (1971). *J. Biol. Chem.* **246**, 2859–2869.
8 Dawson, R.M.C. (1959). *Biochim. Biophys. Acta* **33**, 68–77.
9 Dawson, R.M.C. and Clarke, N. (1972). *Biochem. J.* **127**, 113–118.
10 Dawson, R.M.C., Freinkel, N., Jungalwala, F.B., and Clarke, N. (1971). *Biochem. J.* **122**, 605–607
11 Dawson, R.M.C., Hemington, N., and Irvine, R.F. (1980). *Eur. J. Biochem.* **112**, 33–38.
12 Dawson, R.M.C., Irvine, R.F., and Hirasawa, K. (1982). In *Phospholipids in the Nervous System*, vol. 1, ed. Horrocks, L., pp. 241–249. New York: Raven Press.
13 Dawson, R.M.C., Irvine, R.F., Hirasawa, K., and Hemington, N.L. (1982). *Biochim. Biophys. Acta* **710**, 212–220.

14 De Haas, G.H., Postema, N.M., Nieuwenhuizen, W., and Van Deenen, L.L.M. (1968). *Biochim. Biophys. Acta* **159**, 118–129.
15 Depaoli-Roach, A.A., Gibbs, J.B., and Roach, P.J. (1979). *FEBS Lett.* **105**(2), 321–324.
16 De Torrontegui, G. and Berthet, J. (1966). *Biochim. Biophys. Acta* **116**, 477–481.
17 Fisher, D.B. and Mueller, G.C. (1971). *Biochim. Biophys. Acta* **248**, 434–448.
18 Freinkel, N. (1957). *Endocrinology* **61**, 448–460.
19 Hasegawa-Sasaki, H. and Sasaki, T. (1982). *J. Biochem.* **91**, 463–468.
20 Hirasawa, K. (1983). Ph.D. thesis, University of Cambridge, U.K.
21 Hirasawa, K., Irvine, R.F., and Dawson, R.M.C. (1981). *Biochem. J.* **193**, 607–614.
22 Hirasawa, K., Irvine, R.F., and Dawson, R.M.C. (1981). *Eur. J. Biochem.* **120**, 53–58.
23 Hirasawa, K., Irvine, R.F., and Dawson, R.M.C. (1982). *Biochem. J.* **205**, 437–442.
24 Hirasawa, K., Irvine, R.F., and Dawson, R.M.C. (1982). *Biochem. Biophys. Res. Commun.* **107**, 533–537.
25 Hirasawa, K., Irvine, R.F., and Dawson, R.M.C. (1982). *Biochem. J.* **206**, 675–678.
26 Hofmann, S.L. and Majerus, P.W. (1982). *J. Biol. Chem.* **257**, 6461–6469.
27 Hokin, M.R. and Hokin, L.E. (1953). *J. Biol. Chem.* **203**, 967–977.
28 Irvine, R.F., Hemington, N., and Dawson, R.M.C. (1979). *Eur. J. Biochem.* **99**, 525–530.
29 Irvine, R.F. and Dawson, R.M.C. (1980). *Biochem. Soc. Trans.* **8**, 376–377.
30 Ishizaka, T. and Ishizaka, K. (1983). *Progr. Allergy* **34**, 188–235 (1983).
31 Kemp, P., Hübscher, G., and Hawthorne, J.N. (1961). *Biochem. J.* **79**, 193–200.
32 Keough, K.M.W. and Thompson, W. (1972). *Biochim. Biophys. Acta* **270**, 324–336.
33 Kishimoto, A., Kajikawa, N., Shiota, M., and Nishizuka, Y. (1983). *J. Biol. Chem.* **258**, 1156–1164.
34 Low, M.G. and Finean, J.B. (1976). *Biochem. J.* **154**, 203–208.
35 Masuzawa, Y., Osawa, T., Inoue, K., and Nojima, S. (1973). *Biochim. Biophys. Acta* **326**, 339–344.
36 Meijer, L. and Guerrier, P. (1982). *Biochim. Biophys. Acta* **702**, 143–146.
37 Michell, R.H. (1975). *Biochim. Biophys. Acta* **415**, 81–147.
38 Nishibe, S., Ogawa, M., Murata, A., Nakamura, K., Hatanaka, T., Kambayashi, J., and Kosaki, G. (1983). *Life Sci.* **32**, 1613–1620.
39 Nishizuka, Y. (1983). *Trends Biochem. Sci.* **8**, 13–16.
40 Pickard, M.R. and Hawthorne, J.M. (1978). *J. Neurochem.* **30**, 145–155.
41 Rittenhouse-Simmons, S. (1979). *J. Clin. Invest.* **63**, 580–587.
42 Schultz, K.D., Schultz, K., and Schultz, G. (1977). *Nature* **265**, 750–751.
43 Seals, J.R. and Czech, M.P. (1980). *J. Biol. Chem.* **255**, 6529–6531.
44 Seals, J.R. and Czech, M.P. (1982). *Fed. Proc.* **41**, 2730–2735.
45 Stewart, P.S. and McHennan, D.H. (1974). *J. Biol. Chem.* **249**, 985–993.
46 Takai, Y., Kishimoto, A., Kikkawa, U., Mori, T., and Nishizuka, Y. (1979). *Biochem. Biophys. Res. Commun.* **91**, 1218–1224.
47 Takai, Y., Kaibuchi, K., Sano, K., and Nishizuka, Y. (1982). *J. Biochem.* **91**, 403–406.
48 Takenawa, T. and Nagai, Y. (1981). *J. Biol. Chem.* **256**, 6769–6775.
49 Taverna, R.D. and Hanahan, D.J. (1980). *Biochem. Biophys. Res. Commun.* **94**(2), 652–659.
50 Thompson, W. (1967). *Can. J. Biochem.* **45**, 853–861.

4

CALCIUM AND PHOSPHOLIPID TURNOVER IN SIGNAL TRANSDUCTION; A MECHANISM OF ACTION OF PHORBOL ESTER

USHIO KIKKAWA

Department of Biochemistry, Kobe University School of Medicine, Kobe 650, Japan

It has been generally accepted that various peptide hormones, neurotransmitters and many other biologically active substances provoke inositol phospholipid turnover in their target tissues. The stimulation of most of these receptors immediately mobilizes Ca^{2+} and often increases cGMP but not cAMP. Although inositol phospholipid turnover has sometimes been postulated to be involved in Ca^{2+}-gate opening (*13*), the relationship between the phospholipid turnover and Ca^{2+}-gate opening is still a matter for discussion. The primary products of the receptor-linked breakdown of inositol phospholipids (phosphatidylinositol and its polyphosphates) are identified as diacylglycerol and inositol phosphate (or inositol polyphosphates). However, all attempts to explore the roles of inositol phosphate (inositol polyphosphates) have been unsuccessful. Instead, diacylglycerol is proposed to serve as a membrane fusigen in some exocytotic processes, and phosphatidate therefrom is known as a Ca^{2+} ionophore. Recent studies in this laboratory have uncovered a unique role of diacylglycerol in which it serves as a signal for transmembrane control of protein phosphorylation through activation of a novel protein kinase (*11*, *16*, *19*). This protein kinase has an absolute requirement for Ca^{2+} and phospholipid for its activation, and is referred to as C-kinase.

I. PATHWAY OF SIGNAL TRANSDUCTION

The principal pathway of the signal transduction is schematically given in Fig. 1. At physiologically low concentrations of Ca^{2+}, C-kinase requires diacylglycerol in addition to phospholipid. Diacylglycerol is normally almost absent from membranes, but is transiently produced from inositol phospholipids in a signal-dependent manner. Kinetically, a small quantity of diacylglycerol dramatically increases the affinity of C-kinase for Ca^{2+} and renders this enzyme fully active without a net increase in the Ca^{2+} concentration (*6*, *11*). Among various phospholipids tested phosphatidylserine is indispensable, but other phospholipids show positive or negative cooperativity in this enzyme activation. For instance, in the presence of phosphatidylserine and phosphatidylethanolamine the enzyme is fully active at the 10^{-7} M range of Ca^{2+}, whereas phosphatidylcholine and sphyngomyelin are inhibitory. Thus, the asymmetric distribution of various phospholipids in the lipid bilayer appears to favor the activation of C-kinase (*6*). The enzyme is entirely indifferent to calmodulin, but is profoundly inhibited by many phospholipid-interacting drugs such as trifluoperazine, dibucaine and chlorpromazine, which are all known as calmodulin inhibitors (*14*).

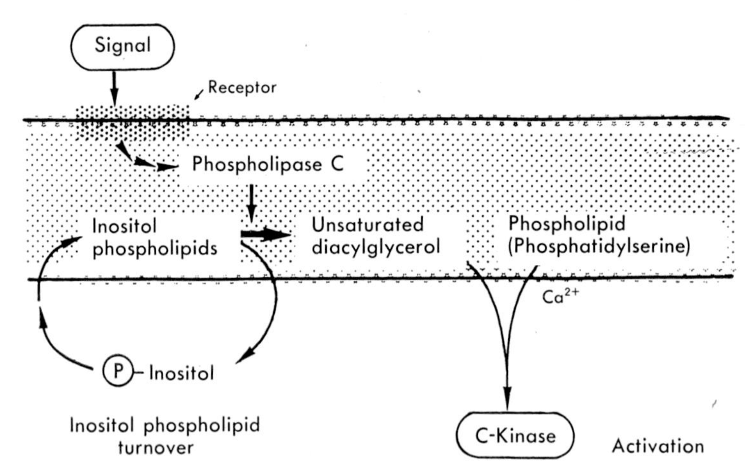

Fig. 1. A proposed pathway of signal transduction for protein phosphorylation.

C-Kinase is distributed ubiquitously in mammalian tissues. In many tissues the activity of this enzyme far exceeds that of cAMP-dependent protein kinase (A-kinase). The enzyme appears to show neither tissue nor species specificity in its kinetic and catalytic properties, and was recently purified to homogeneity from rat brain (*10*) and also by Kuo and his coworkers (*22*) from bovine heart muscle. The enzyme is a single polypeptide with no subunit structure; molecular weight is estimated to be about 77,000 and it shows a broad substrate specificity. C-Kinase and A-kinase can often use the same phosphate acceptor proteins, but analysis *in vitro* of the phosphorylation sites indicates that these kinases show distinctly different catalytic properties; each appears to recognize the respective seryl and threonyl residues in common substrate proteins. C-Kinase also phosphorylates itself, but the significance of this autophosphorylation is not known. The enzyme does not react with its own tyrosyl residues or those in any substrate proteins tested.

II. EVIDENCE FOR RECEPTOR-LINKED PROTEIN PHOSPHORYLATION

Platelets are frequently used as a model system for hormone action since many agonists and antagonists for the aggregation and release reactions are known. Thus, the experimental evidence presented in this review has been obtained mostly with human platelets unless otherwise specified. When stimulated by thrombin, collagen or platelet-activating factor (PAF), two endogenous platelet proteins with approximate molecular weights of 40,000 (40K protein) and 20,000 (20K protein) are heavily phosphorylated, and this phosphorylation reaction always accompanies the release of dense body constituents such as serotonin (*3*, *9*, *12*). 20K protein has been identified as myosin light chain, and a calmodulin-dependent protein kinase is responsible for this reaction (*4*). Although the function of 40K protein remains unknown, this protein isolated from platelets serves as a substrate for C-kinase. Fingerprint analysis of the radioactive 40K protein phosphorylated *in vivo* and *in vitro* suggests that C-kinase is responsible for the protein phosphorylation that is observed during platelet activation. In fact, when human platelets are stimulated, diacylglycerol containing arachidonate is rapidly produced with concomitant disappearance of phosphatidylinositol, and this reaction is always associated with 40K protein phosphorylation (*5, 9, 18*). Under appropriate conditions, synthetic diacylglycerol such as 1-oleoyl-2-acetyl-glycerol added to platelets is able to induce 40K protein phosphorylation directly (*8*). This exogenous diacylglycerol is rapidly converted *in situ* to phosphatidate, 1-oleoyl-2-acetyl-3-phosphoryl-glycerol, probably through the action of diacylglycerol kinase.

Although the possible roles of C-kinase have been most extensively studied with human platelets, evidence available at present suggests that this enzyme plays a key role in many other tissues in signal transduction for protein phosphorylation. For example, rat neutrophils are activated by fMet-Leu-Phe, and release lysosomal enzymes. Rat mast cells are activated by concanavalin A, and release histamine. When these cells are incubated with the synthetic diacylglycerol, the release reactions are induced. It is known that the extracellular messengers mentioned above elicit inositol phospholipid turnover in their respective target cells.

III. SYNERGISTIC ROLES OF PROTEIN PHOSPHORYLATION AND CALCIUM MOBILIZATION

Although the activation of C-kinase absolutely requires Ca^{2+}, synthetic diacylglycerol directly added to intact platelets causes the full activation of this enzyme in the absence of Ca^{2+} mobilization. This is presumably due to the fact that diacylglycerol dramatically increases the affinity of C-kinase for Ca^{2+}, and renders this enzyme fully active with-

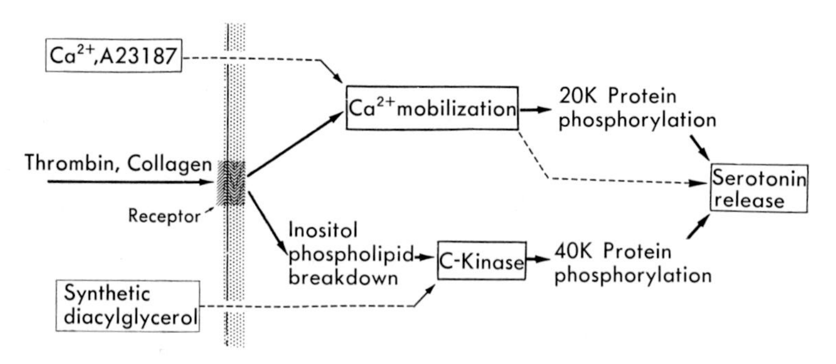

Fig. 2. Independent induction of Ca^{2+} mobilization and protein phosphorylation by Ca^{2+} ionophore and synthetic diacylglycerol. Taken from ref. *17*.

out a net increase in the cytosolic Ca^{2+} concentration as mentioned above. However, the activation of C-kinase appears to be a prerequisite though not a sufficient requirement for the release reaction, and the addition of synthetic diacylglycerol alone does not result in the full response of platelets to release serotonin. As schematically shown in Fig. 2, it is experimentally possible to induce Ca^{2+} mobilization and C-kinase activation independently by the addition of Ca^{2+} ionophore and synthetic diacylglycerol, respectively. In platelets, the mobilization of Ca^{2+} may be easily monitored by quantitating the phosphorylation of 20K protein (myosin light chain) which absolutely depends on calmodulin. When platelets are incubated with 1-oleoyl-2-acetyl-glycerol, the release of serotonin is markedly enhanced by the addition of ionophore A23187, and the full physiological response may be observed in the presence of both diacylglycerol and ionophore A23187 (*8*, *17*). In contrast, 40K protein phosphorylation is not significantly affected by the same treatment, and this protein is almost fully phosphorylated by the exogenous addition of diacylglycerol alone, just as it is by natural extracellular messengers such as thrombin, collagen and PAF. The Ca^{2+} ionophore alone does not induce endogenous diacylglycerol formation and 40K protein phosphorylation, or cause release of serotonin at the concentration employed for the experiment. However, this ionophore at more than the micromolar range is responsible for both 40K protein phosphorylation and serotonin release, probably due to enhancement of non-specific degradation of phospholipids and/or to activation of C-kinase in the absence of diacylglycerol by a large increase of Ca^{2+} (*8*). Likewise, this synthetic diacylglycerol alone at more than 0.1 mg/ml causes release of a significant amount of serotonin. The exact reason for this enhanced serotonin release is not known, but it is possible that diacylglycerol or phosphatidate therefrom may act as membrane fusigen or Ca^{2+} ionophore. Ca^{2+} may play diverse roles in this physiological process. Nevertheless, the results briefly outlined above seem to indicate that both Ca^{2+} mobilization and protein phosphorylation are indispensable for full cellular response to extracellular messengers.

IV. CYCLIC NUCLEOTIDES AS NEGATIVE RATHER THAN POSITIVE MESSENGERS

Most tissues seem to possess at least two major classes of receptors for controlling cellular functions and proliferation, although there is a dramatic variation and heterogeneity from tissue to tissue. One class of receptors is related to cAMP, and the other generally induces inositol phospholipid breakdown as well as immediate Ca^{2+} mobilization as mentioned above. In addition, the stimulation of the latter class of receptors often produces cGMP. In *bidirectional control systems* for most tissues such as platelets, the receptors that induce inositol phospholipid turnover normally promote activation of cellular functions and proliferation, whereas the receptors that produce cAMP usually antagonize such activation. With human platelets it has been shown that phosphatidylinositol breakdown, diacylglycerol formation, 40K protein phosphorylation, and serotonin release are all inhibited in a parallel manner by prostaglandin E_1 as well as dibutyryl cAMP (*7*, *21*).

Although cGMP has been repeatedly shown to be increased by stimulation of the receptors that induce inositol phospholipid turnover, its definitive role in biological regulation is not yet well understood. It remains a puzzle that cGMP-dependent protein kinase (G-kinase)

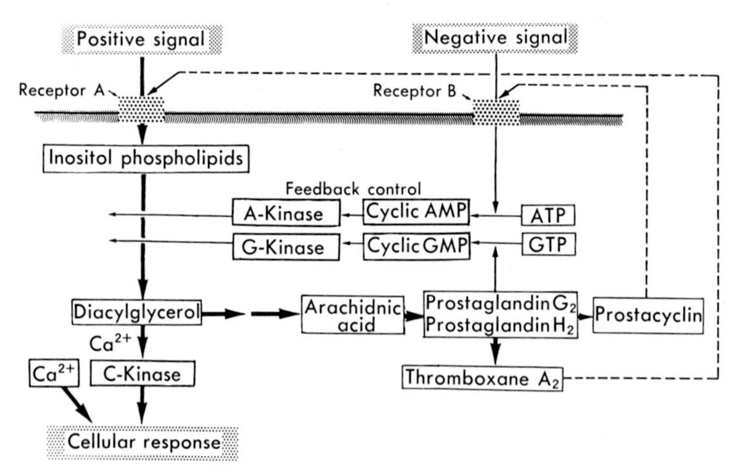

Fig. 3. Phospholipid degradation in signal transduction in platelets as model system. Taken from refs. *16*, *21*.

shows very similar, if not identical, catalytic properties to those of A-kinase (*2*). In human platelets cGMP has been suggested to act as a negative rather than a positive messenger and to constitute an immediate feedback control that prevents over-response (*20*). In fact, either sodium nitroprusside or 8-bromo-cGMP inhibits the thrombin-induced phosphatidylinositol breakdown, diacylglycerol formation, 40K protein phosphorylation and serotonin release in a parallel manner.

These results are compatible with the supposition that cAMP and cGMP do not antagonize each other but similarly inhibit the phospholipid degradation in the activated cell membrane. Although the mechanism of cGMP accumulation is not clear at present, it is possible that arachidonate peroxide and prostaglandin endoperoxides may serve as activators for guanylate cyclase. Such a possible receptor-linked phospholipid metabolic cascade for signal transduction is schematically given in Fig. 3.

V. A MECHANISM OF ACTION OF PHORBOL ESTER

TPA (12-*O*-tetradecanoylphorbol-13-acetate) can substitute for diacylglycerol at extremely low concentrations, and directly activates C-kinase *in vitro* in the presence of both Ca^{2+} and phospholipid (*1*). TPA alone shows no effect. Other protein kinases tested are not susceptible to TPA. Again, the experiments with human platelets indicate that this tumor promoter induces 40K protein phosphorylation and serotonin release, but not diacylglycerol formation. In addition, various phorbol derivatives showing tumor-promoting activity can activate C-kinase, and the structural requirements of phorbol-related diterpenes for tumor promotion appear similar to those for this enzyme activation. A number of kinetic studies seem to indicate that these tumor promoters have specific cell surface receptors that are present in a wide variety of tissues. It has been proposed that one of these receptors may be the C-kinase phospholipid complex (*1*, *16*). A recent report presented by Niedel *et al.* (*15*) favors this proposal. Under physiological conditions, diacylglycerol is produced only transiently and disappears within a minute at most when receptors are stimulated by ligands such as neurotransmitter or hormone. In contrast, the tumor promoter intercalates into the phospholipid bilayer over a prolonged period since it is metabolized only very slowly. The cell thus tends to proliferate since C-kinase is kept active despite the feedback control by cyclic nucleotides.

VI. CONCLUSION

Although the experimental support briefly described above has resulted from studies with a limited number of specific tissues, it is suggestive that C-kinase plays a crucial role in transmitting information from many extracellular messengers. However, details of the target proteins of this enzyme in most tissues remain to be clarified. The function of 40K protein in platelets is to be explored further. In addition, the event occurring immediately after stimulation of the receptors must also be clarified. Phospholipases C in mammalian tissues so far described are specific for inositol phospholipids and do not appear to be rate-limiting in any type of tissue. Perhaps under normal conditions inositol phospholipids are present in forms which cannot be attacked by this class of enzymes; one serious barrier lies in the difficulty of understanding the lipid-enzyme interaction. Also, evidence still appears insufficient to discuss the relationship between inositol phospholipid breakdown and Ca^{2+} mobilization. However, evidence presented above seems to indicate that protein phosphorylation and Ca^{2+} mobilization are equally essential and act synergistically to elicit full physiological cellular response. Obviously, each of these events may play a diverse role in the activation of cellular functions and proliferation. However, the underlying principle proposed in this review may be modified by further knowledge that will expand rapidly in the next few years.

SUMMARY

Inositol phospholipid turnover provoked by a wide variety of extracellular messengers such as hormones, neurotransmitters, secretagogues, growth factors and many other biologically active substances appears to be directly coupled to protein phosphorylation through activation of a novel protein kinase (C-kinase). This protein kinase activation is a prerequisite for the transmembrane control of cellular functions and proliferation, and evidence is presented that the receptor-linked protein phosphorylation and Ca^{2+} mobilization act synergistically to elicit full physiological cellular response. In many tissues cAMP and cGMP do not antagonize each other but similarly block the phospholipid degradation and thereby counteract the signal transduction. Tumor-promoting phorbol ester intercalates into membrane phospholipid bilayer and directly activates C-kinase without inducing inositol phospholipid turnover.

Acknowledgments
The author wishes to thank Prof. Y. Nishizuka and Associate Prof. Y. Takai for valuable discussion and support in preparing this review. He also expresses indebtedness to his collaborators K. Kaibuchi, K. Sano, M. Castagna, J. Yamanishi, R. Miyake and Y. Tanaka. This investigation has been supported in part by research grants from the Ministry of Education, Science and Culture, the Ministry of Health and Welfare, and the Science and Technology Agency, Japan.

REFERENCES

1 Castagna, M., Takai, Y., Kaibuchi, K., Sano, K., Kikkawa, U., and Nishizuka, Y. (1982) *J. Biol. Chem.* **257**, 7847–7851.
2 Hashimoto, E., Takeda, M., Nishizuka, Y., Hamana, K., and Iwai, K. (1976). *J. Biol. Chem.* **251**, 6287–6293.
3 Haslam, R.J. and Lynham, J.A. (1977). *Biochem. Biophys. Res. Commun.* **77**, 714–722.
4 Hathaway, D.R. and Adelstein, R.S. (1979). *Proc. Natl. Acad. Sci. U.S.* **76**, 1653–1657.
5 Ieyasu, H., Takai, Y., Kaibuchi, K., Sawamura, M., and Nishizuka, Y. (1982). *Biochem. Biophys. Res. Commun.* **108**, 1701–1708.
6 Kaibuchi, K., Takai, Y., and Nishizuka, Y. (1981). *J. Biol. Chem.* **256**, 7146–7149.
7 Kaibuchi, K., Takai, Y., Ogawa, Y., Kimura, S., Nishizuka, Y., Nakamura, T., Tomomura, A., and Ichihara, A. (1982). *Biochem. Biophys. Res. Commun.* **104**, 105–112.
8 Kaibuchi, K., Takai, Y., Sawamura, M., Hoshijima, M., Fujikura, T., and Nishizuka, Y. (1983). *J. Biol. Chem.* **258**, 6701–6704.
9 Kawahara, Y., Takai, Y., Minakuchi, R., Sano, K., and Nishizuka, Y. (1980). *Biochem. Biophys. Res. Commun.* **97**, 309–317.
10 Kikkawa, U., Takai, Y., Minakuchi, R., Inohara, S., and Nishizuka, Y. (1982). *J. Biol. Chem.* **257**, 13341–13348.
11 Kishimoto, A., Takai, Y., Mori, T., Kikkawa, U., and Nishizuka, Y. (1980). *J. Biol. Chem.* **255**, 2273–2276.
12 Lyons, R.M., Stanford, N., and Majerus, P.W. (1975). *J. Clin. Invest.* **56**, 924–936.
13 Michell, R.H. (1975). *Biochim. Biophys. Acta* **415**, 81–147.
14 Mori, T., Takai, Y., Minakuchi, R., Yu, B., and Nishizuka, Y. (1980). *J. Biol. Chem.* **255**, 8378–8380.
15 Niedel, J.E., Kuhn, L.J., and Vandenbark, C.R. (1983). *Proc. Natl. Acad. Sci. U.S.* **80**, 36–40.

16 Nishizuka, Y. (1983). *Trends Biochem. Sci.* **8**, 13–16.
17 Nishizuka, Y. (1983). *Phil. Trans. R. Soc. Lond.* **B302**, 101–112.
18 Sano, K., Takai, Y., Yamanishi, J., and Nishizuka, Y. (1983). *J. Biol. Chem.* **258**, 2010–2013.
19 Takai, Y., Kishimoto, A., Iwasa, Y., Kawahara, Y., Mori, T., and Nishizuka, Y. (1979). *J. Biol. Chem.* **254**, 3692–3695.
20 Takai, Y., Kaibuchi, K., Matsubara, T., and Nishizuka, Y. (1981). *Biochem. Biophys. Res. Commun.* **101**, 61–67.
21 Takai, Y., Kaibuchi, K., Sano, K., and Nishizuka, Y. (1982). *J. Biochem.* **91**, 403–406.
22 Wise, B.C., Raynor, R.L., and Kuo, J.F. (1982). *J. Biol. Chem.* **257**, 8481–8488.

5

TRANSMEMBRANE SIGNALLING IN AREAS OF CELL CONTACT

BENJAMIN GEIGER

Department of Chemical Immunology, The Weizmann Institute of Science, Rehovot 76100, Israel

Cell contacts (junctions) are believed to play key roles in determining a variety of cellular activities and interactions. Contact with neighboring cells or extracellular matrices has an obvious effect on cell morphology and motility and on the assembly of individual cells into intact tissues. Moreover, many recent studies have shown that the normal physiology of cells is highly dependent upon cell contact. This includes both cell growth and expression of differentiated cellular functions. Structural studies using either electron microscopy or immunocytochemical localization of cytoskeletal elements indicated that at least two types of cell junctions are associated at their endofacial surfaces with cytoskeletal filaments. *Adhaerens* type junctions are associated with actin bundles and desmosomes with intermediate filaments, mostly of the prekeratin family.

The well-documented transmembrane associations with the cytoskeleton together with the physiological data mentioned above have suggested the possibility that signals important for regulation of cell behavior are transmitted across the membrane in these junctions (*6–8*). In our work we have focused mostly on the *adhaerens* junction and some ideas regarding its molecular architecture and biogenesis will be discussed here.

The *adhaerens* junctions, as we define them, are sites of contact with neighboring cells or with extracellular matrix (ECM) which are associated at their cytoplasmic surfaces with actin and vinculin (*5–12*). In a series of studies over the last few years we have identified several cellular contacts which fit this definition. These include the intercellular contacts in polarized epithelium (*zonula adhaerens*) and in cardiac myocytes (*fascia adhaerens*), dense plaques of smooth muscle, focal contacts and intercellular attachments of cultured cells, *etc.* All these sites were shown to be attached to actin and to contain vinculin close to or on the endofacial surfaces of the junctional membrane.

On the basis of these early observations we have proposed a working model for the molecular steps in the assembly of these junctions. As will be detailed later we assumed that small local contacts grow, leading to the accumulation of certain membrane receptors in the area of contact. These "patches" induce the local accumulation of vinculin at the interior. Vinculin subsequently assembles actin into small premordial bundles that eventually grow into stress fibers. We have thus postulated that vinculin plays a key role in the induction of actin assembly and some of these aspects will be outlined here. In this paper I will present little new data but will discuss our current views on junction formation and rearrangement and on the nature of the transmembrane signals transmitted in the sites of contact in a rather general review.

I. VINCULIN, A PUTATIVE LINKER OF ACTIN TO THE JUNCTIONAL MEMBRANE

We initially isolated vinculin several years ago as a side product during the purification of α-actinin from chicken gizzard smooth muscle (*5*). The protein exists as a monomer in aqueous buffers over a wide pH range with an apparent molecular weight of 130,000; it is non-glycosylated and strictly intracellular. Two dimensional gel electrophoretic analysis indicated that in cells and tissues vinculin exists as a family of several isoforms which vary in their pI by up to 0.5 pH unit (*7*). The acidic form, amounting to about 5% of the total protein, is phosphorylated mostly on seryl residues. In addition to the major

130,000 dalton form in several tissues (such as gizzard, heart, intestine and skin) a related protein with higher molecular weight was detected (*20* and Volberg and Geiger, unpublished) which reacted with anti-pure vinculin and had a peptide map pattern similar to that of the 130,000 dalton protein. Differential extraction experiments suggested that the high molecular weight variant, denoted metavinculin, seems to require both high salt and detergent for an efficient extraction and it was proposed that this form might be tightly associated with the cell membrane.

In a series of immunoelectron microscopic studies we have shown that vinculin is closely distributed in the vicinity of the junctional membrane (closer than α-actinin and other actin-associated proteins) though we could not exclude, by this type of experiment, the possibility that its linkage to the membrane may be mediated by other peripheral protein(s) (*11*).

Another related property of vinculin is its special mode of interaction with actin. It was shown that vinculin binds specifically to F-actin *in vitro* and that its addition, even at low molar ratios (down to $\sim$1:1000) to actin causes a marked reduction in the viscosity of the F-actin solution (*4, 14, 15, 21*). This effect was interpreted as indicating that either vinculin causes the bundling of actin filaments or that it caps or severs them. Electron microscopic observations suggested that vinculin was indeed capable of bundling actin in solution. In conclusion, the microscopic data, together with the biochemical characterization of vinculin and its interaction with microfilaments pointed to its close involvement in the formation of *adhaerens* junctions.

II. STAGES IN THE BIOGENESIS OF *ADHAERENS* JUNCTIONS: FACTS AND THEORY

The biochemical data on vinculin and its properties as well as the immunolocalization of the protein provided a rather static view of the junction and the associated cytoskeleton and failed to elucidate the mode of action of vinculin during junction formation, maintenance, modulation and reversal. *Adhaerens* junctions such as focal contacts are known to be rather dynamic structures. During locomotion cells are known to form anterior contacts and to detach the posterior ones. Cells undergoing mitosis reverse most of their substrate contacts and reestablish them only when the two daughter cells respread. Moreover, visualization of cell-substrate contacts by interference-reflection microscopy (IRM) indicated that individual attachment plaques can grow, become reoriented, decrease in size, *etc.*

The exact spatial and temporal chain of molecular events leading to the establishment of nascent focal contacts was not known at the time we started our studies and is poorly understood even today. This local process is undoubtedly very complex and could involve a large number of factors including ECM and membrane components, vinculin and related proteins, state of organization of the actin, local ion fluxes, membrane potential, *etc.* At the present time we cannot include all these factors and possibly many others into one combined and coherent model for junction biogenesis. We nevertheless, were able to come up with a minimal working model for junction formation which could (at least partly) be tested experimentally.

This model (for graphic representation, see refs. *6* and *9*) suggests that cells contain specific "contact receptors" or even a family of such molecules on their surface. When the cells are suspended these receptors may freely diffuse laterally at the plane of the membrane. Vinculin at that stage is mostly soluble in the cytoplasm and actin forms a cortical mesh of randomly oriented filaments. Upon formation of small contacts the entire system is perturbed and a series of transmembrane nucleation events occur. First, the "contact receptors" aggregate around the initial site of contact in a positively cooperative process and become immobilized. This lateral clustering provides a transmembrane signal and induces the formation of vinculin binding sites at the cytoplasmic faces of the membrane. This process may be direct (the "contact receptors" might have a cytoplasmic domain which, when aggregated, binds vinculin) or indirect, through other integral or peripheral proteins. The soluble vinculin binds to the contact area and forms a patch on the membrane. This area then becomes a nucleation center for the assembly or actin bundles, the formation of which may involve either polar local polymerization of G actin or collection and reorientation of preexisting F-actin filaments. This new membrane-bound bundle may subsequently grow due to the action of different actin binding proteins and eventually develop into a stress fiber.

This model, put forward a few years ago, was entirely hypothetical and many of its aspects and predictions have still not been tested directly. Nevertheless, several studies carried out by us and others on the dynamics of contact formation and the rearrangements of the bound actin support its general outline.

From the theoretical point of view we were nonetheless intrigued by the highly ordered pattern of substrate contacts during the early stages of cell spreading. Initially the cells tend to form broad and uniform substrate contacts which develop within a few minutes into a series of radial focal contacts with a rather uniform periodicity. This early formation of periodic structures was assumed to result from the growth of local perturbations to the uniform state which drastically decreases the lateral diffusion of the "contact receptors". Without making any presumptions on the involvement of the cytoskeleton we have offered the theory that periodic contacts may evolve and their "wavelength" may depend primarily on the "adhesiveness" of the substrate and the lateral diffusion coefficients of the free and anchored receptors (*19*).

III. DYNAMICS OF *ADHAERENS* JUNCTION COMPONENTS

One of the major predictions of the model outlined above is that all elements of focal contact are dynamic and that some steady state equilibrium exists between the junctional and extrajunctional pools. When the equilibrium is perturbed in one way or another the contact and bound cytoskeleton may grow or decrease in size.

Direct proof of the dynamic nature of focal contacts was obtained by the use of the fluorescence photobleaching recovery (FPR) technique. This method (see ref. *18*) uses fluorescent probes which are introduced to the cell membrane or to the cytoplasm. The fluorescence in small areas is then bleached by a focused laser beam and the rate of fluorescence recovery measured. Since the photobleaching is itself irreversible, the rate of recovery of fluorescence depends on the diffusion coefficient of the probe which can easily be calculated.

To determine membrane dynamics we have introduced into the cell membrane a lipid probe (WW 591) or have labeled surface proteins with the NH_2-group specific non-penetrating fluorophore lissamine rhodamine sulfonyl chloride. The cells were then allowed to adhere and their ventral (substrate attached) membranes isolated in an intact state by the $ZnCl_2$ method of Avnur and Geiger (*1*). Focal contacts were then localized by IRM and FPR measurements performed within and outside the site of contact (for complete results see ref. *8*). In short, we found that both proteins and lipids were mobile within focal contacts (though at a reduced rate). However, while all the lipid probe was free to diffuse laterally a significant proportion ($\sim 50\%$) of the fluorescence of proteins was apparently immobile at the time scale of FPR experiments ($D \leq 10^{-12}$ cm^2/sec). The existence of both free and immobile microdomains in this *adhaerens* junctions is certainly in line with the proposed model.

Analysis of the dynamic properties of the cytoskeletal domains of the focal contact by a similar technique required the microinjection of the relevant cytoskeletal proteins (actin, vinculin, and α-actinin, labeled with rhodamine or fluorescein) into living cells. Shortly after injection the labeled proteins became associated with the microfilament system of the host cell. Actin was found in stress fibers, focal contact, and leading lamella. α-Actinin was enriched in focal contacts and displayed periodic striations along the stress fibers and vinculin was associated specifically with focal contacts. FPR measurements on living cells indicated that each protein maintained two cellular pools, a diffusible pool and an "anchored pool". Actin was primarily associated with the interfibrillary space while α-actinin was present predominantly on stress fibers and focal contacts. "Immobile" vinculin was detected in focal contacts only. The FPR data suggested that most, if not all, the junction-bound cytoskeletal elements were not free to diffuse. We, however, found that these components were not entirely immobile; long (few minutes) tracing of bleached areas or stress fibers or focal contacts indicated that the fluorescence slowly recovers (*9*, *16*). The rate of recovery was not affected by changing the size of the bleaching beam, suggesting that exchange between the two pools was the predominant mechanism.

The FPR experiments have established that there is indeed a continuous exchange of junctional components between the diffusible and "organized" pools and thus a shift, even a small one, in this equilibrium may either cause growth of the contact area or decrease its size. These experiments provided some ideas regarding the overall dynamics of the contact area and suggested that each component moves independently. The FPR approach could not, however, provide information regarding the fine molecular interactions and interconnections of these elements and the latter problem was thus studied using a different approach which will be briefly discussed in the following chapter.

IV. ASSOCIATION OF VINCULIN TO THE JUNCTIONAL MEMBRANE REQUIRES CONTINUOUS MAINTENANCE OF CONTACT BUT IS ACTIN INDEPENDENT

The electron microscopic data suggested that vinculin was closely associated with the plasma membrane, closer than α-actinin and tropomyosin. This approach, however, did not have sufficient resolution to determine whether vinculin was indeed directly associated with the membrane or whether its primary attachment was to actin. As pointed out above, *in vitro* studies on the effects of vinculin on the viscosity of actin indicated that the two can specifically interact and that vinculin can induce actin bundling.

To investigate indirectly the nature of the binding sites for endogenous and exogenous vinculin we prepared ventral membranes by the $ZnCl_2$ method and selectively removed actin from them using the F-actin severing protein fragmin from physarum (*13*). This protein rapidly severs actin in a Ca^{2+}-dependent manner. We have shown that vinculin and α-actinin were both associated with the termini of actin bundles in the untreated membranes. After fragmin treatment both actin and α-actinin (as well as tropomyosin, myosin, and filamin) were removed while vinculin was essentially unaffected. Moreover, binding of exogenous fluorescently labeled vinculin to the membrane fragments was not affected by this treatment. It was thus concluded that the binding of vinculin to focal contacts was largely actin independent in contrast to the other proteins tested (*2*).

An interesting related observation was that the "decoration" with fluorescent vinculin could not be efficiently inhibited by excess unlabeled vinculin. This strongly suggested that vinculin may undergo self aggregation on the membrane.

Is vinculin directly associated with an integral membrane component? The answer to that question is not yet clear. Recent studies have indicated that another protein with a molecular weight of 215,000 called "talin" (*3*) has a very similar distribution to that of vinculin. Preliminary results indicated that talin can bind vinculin (*17*). This, together with the binding of the latter to actin and to itself, suggests that the chain of interactions may involve (in this order) elements in the plasma membrane-talin-vinculin (in several layers)-actin and associated proteins. These is still no direct proof for this model and considerable effort will be necessary to test its various aspects.

We were nevertheless able to show that the attachment of vinculin to the junctional membrane depends on the continuous maintenance of the cell contact. We have shown that after removal of extracellular Ca^{2+} ions from cultured epithelial sheets of Madin-Darby bovine kidney (MDBK) cells the subapical belt of actin, α-actinin and vinculin dissociated from the membrane and contracted together towards the cell center. Whether talin is also removed by this treatment is not yet known.

SUMMARY

The studies briefly summarized here provide some ideas regarding the general rules underlying the contact-induced assembly of the microfilament system. They undoubtedly leave many questions unanswered: Are all *adhaerens* junctions composed of the same set of proteins? What are the membrane components which mediate the cell-substrate or intercellular contact? What are the spatial interrelationships between vinculin and talin? Are there additional cytoskeletal elements involved in the anchorage of microfilaments? Are the "structural" transmembrane signals transmitted in *adhaerens* junctions related to the contact-dependent regulation of cell growth and differentiation, and how? Hopefully, answers to some of these questions will be available in the near future and will shed some light on the contact-induced regulation of cell activity.

Note Added in Proof: Recent results on the microinjection of vinculin and the decoration of ventral membranes with the labeled protein were described by T.E. Kreis.

REFERENCES

1 Avnur, Z. and Geiger, B. (1981). *J. Mol. Biol.* **153**, 361–379.
2 Avnur, Z., Small, J.V., and Geiger, B. (1983). *J. Cell Biol.* **96**, 1622–1630.
3 Burridge, K. and Connell, L. (1983). *J. Cell Biol.* **97**, 359–367.
4 Burridge, K. and Feramisco, J.R. (1982). *Cold Spring Harbor Symp. Quant. Biol.* **46**, 587–593.
5 Geiger, B. (1979). *Cell* **18**, 193–205.
6 Geiger, B. (1982). *Cold Spring Harbor Symp. Quant. Biol.* **46**, 671–682.
7 Geiger, B. (1982). *J. Mol. Biol.* **159**, 685–701.
8 Geiger, B., Avnur, Z., and Schlessinger, J. (1982). *J. Cell Biol.* **93**, 495–500.
9 Geiger, B., Avnur, Z., Kreis, T.E., and Schlessinger, J. In *Cell Muscle Motility*, ed. Shay, J., New York: Plenum Press, in press.
10 Geiger, B., Tokuyasu, K.T., Dutton, A.H., and Singer, S.J. (1980). *Proc. Natl. Acad. Sci. U.S.* **77**, 4127–4131.
11 Geiger, B., Dutton, A.H., Tokuyasu, K.T., and Singer, S.J. (1981). *J. Cell Biol.* **91**, 614–628.
12 Geiger, B., Schmid, E., and Franke, W.W. (1983). *Differentiation* **23**, 189–205.
13 Hinssen, H. (1981). *Eur. J. Cell Biol.* **23**, 225–233.
14 Jockusch, B.M. and Isenberg, G. (1981). *Proc. Natl. Acad. Sci. U.S.* **78**, 3005–3010.
15 Jockusch, B.M. and Isenberg, G. (1982). *Cold Spring Harbor Symp. Quant. Biol.* **46**, 613–622.
16 Kreis, T.E., Geiger, B., and Schlessinger, J. (1982). *Cell* **29**, 835–845.
17 Mangeat, P. and Burridge, K. (1984). *J. Cell Biol.* in press.
18 Schlessinger, J. and Elson, E. (1982). *Method Experim. Phys.* **20**, 197–227.
19 Segel, L., Volk, T., and Geiger, B. *J. Cell Biophys.* in press.
20 Silicano, J. and Craig, S.W. (1982). *Nature* **300**, 533–535.
21 Wilkins, J.A. and Lin, S. (1982). *Cell* **28**, 83–91.

6

THE CYTOSKELETON UNDERLYING THE AXOLEMMA

SHOICHIRO TSUKITA AND SACHIKO TSUKITA

Department of Anatomy, Faculty of Medicine, University of Tokyo, Tokyo 113, Japan

It is now widely accepted that the cytoskeleton is somewhat directly involved in the process of "transmembrane signaling and sensation." Since the axonal cytoskeleton is structurally rather simple in its variety of components and the axolemma is an excitable membrane, the axon seems to offer an advantageous model for the studies of the interaction between the cytoskeleton and the excitable membrane. Actually, the axonal cytoskeleton has often been discussed recently in the light of its roles in maintaining the excitability of the axolemma (*3*, *10*, *11*). In order to understand these physiological roles of the axonal cytoskeleton at the molecular level, correlative morphological and biochemical studies of the molecular organization of the axonal cytoskeleton are required. In this article, we first describe the three-dimensional organization of the axonal cytoskeleton using a rapid-freeze, deep-etch, rotary-shadowing method (*21*). Secondly, we survey our recent studies of two different types of high molecular weight proteins, calspectin (*8*, *18*) and 260 K protein (*19*), which might play an important role in connecting the axonal cytoskeleton to the axolemma.

I. THREE-DIMENSIONAL ORGANIZATION OF THE AXONAL CYTOSKELETON

Recent technical improvement in the deep-etch, rotary-shadowing replica method aided by rapid freezing using liquid helium has made possible direct electron microscopic analysis of fresh biological specimens without any chemical fixation (*6*, *24*). To further advantage, the replicas lend themselves well to stereoscopic observation at high resolution. Therefore, at first, using this new technique, we have attempted to analyze the three-dimensional organization of the axonal cytoskeleton in rat myelinated axons (*21*).

In this study, we used rat trigeminal nerves. The nerve was dissected out and cut longitudinally into two pieces by a razor blade. Then it was rapidly frozen with a newly-developed freezing device, RF-10 (Eiko Engineering, Mito, Japan) (*23*): the nerve's newly exposed surface was touched against a pure copper block cooled to 4°K by liquid helium (*20*, *21*), basically according to the procedure of Heuser and Salpeter (*6*). The frozen tissue was brought into a cryo-kit on a Sorvall MT-2 ultramicrotome and cut with a glass knife at −120°C to remove the most superficial 10 μm-thick layer of the metal-contact tissue surface. The tissue was then transferred to a freeze-etching device, FD-3 (Eiko Engineering, Mito, Japan), in which the exposed tissue surface was deeply etched and rotary-shadowed with platinum-carbon to obtain replicas.

The characteristic feature of such freeze-etch replicas was that longitudinally oriented microtubules and neurofilaments were linked with interconnecting strands to form a dense, three-dimensional latticework (Fig. 1). These strands can be considered to correspond to the wispy filamentous structures or microtrabecular structures seen in chemically fixed axons (*2*). The microtubules were mixed non-randomly with neurofilaments in the axoplasm, forming the domains (territories) occupied by microtubules. There seemed to be some differences in the arrangement of such interconnecting strands between microtubule territories and neurofilament territories. In the microtubule territories, the interconnecting strands also joined to form complicated networks, while in the neurofilament territories, the strands extended

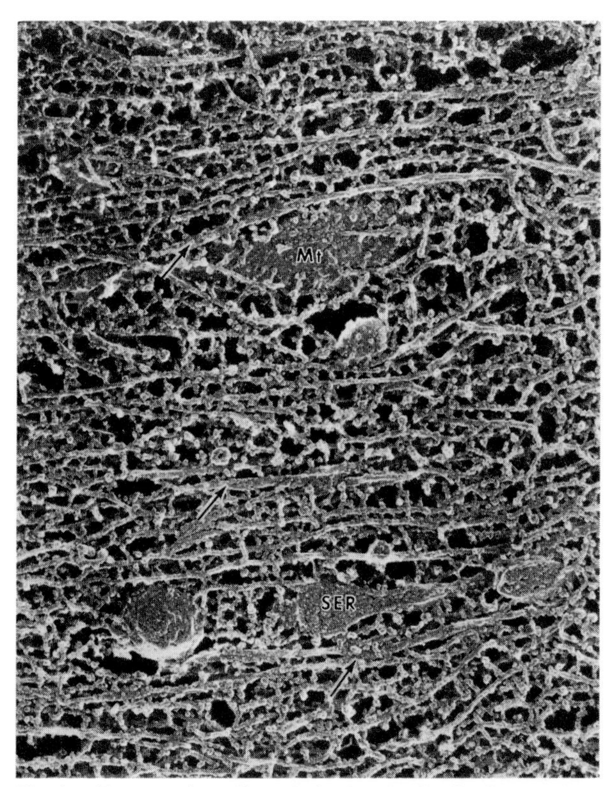

Fig. 1. Freeze-etch replica of the longitudinally fractured myelinated axon in an unfixed trigeminal nerve. The axoplasm is seen to be a densely packed fibrous network, which mainly comprises microtubules (arrows), neurofilaments, and their interconnecting strands. Membranous organelles such as mitochondria (Mt) and smooth endoplasmic reticulum (SER) are distributed within the cytoskeletal network. Note the interconnecting strands between membranous organelles and cytoskeletal structures. ×80,000.

laterally from a neurofilament to merely connect adjacent neurofilaments showing a ladder-like appearance. Membranous organelles such as axoplasmic reticulum and mitochondria were distributed within the cytoskeletal network in the axoplasm (Figs. 1 and 2a). Careful observations at higher magnification revealed that the interconnecting strands extending from neurofilaments and microtubules were directly attached to the cytoplasmic surface of the axoplasmic reticulum and mitochondria. In many replicas, these membranous organelles showed a more intimate spatial relationship with microtubules than with the neurofilaments. These interconnecting strands between microtubules and

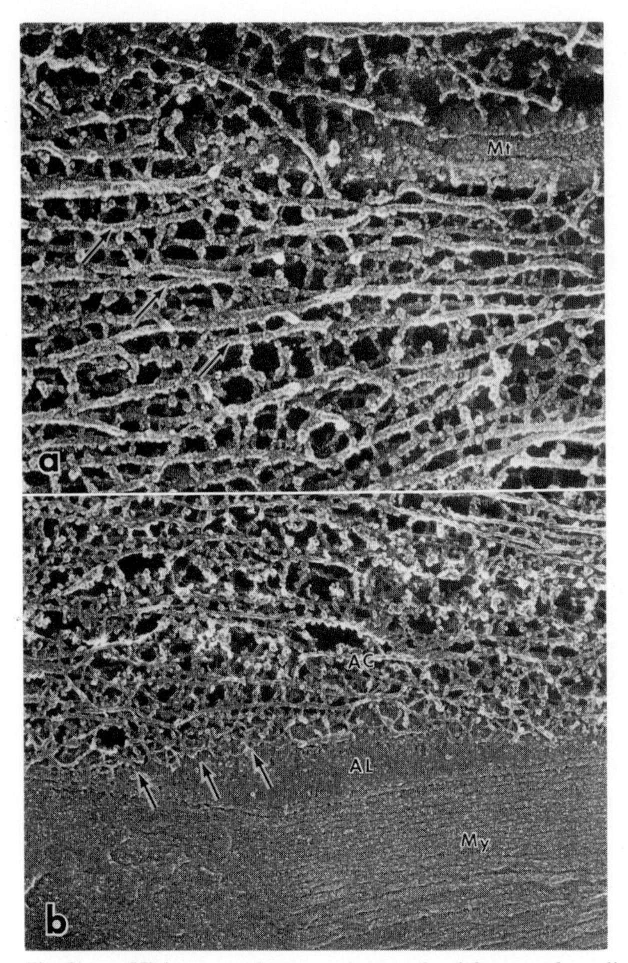

Fig. 2. a, High-power electron micrograph of freeze-etch replica of the longitudinally fractured myelinated axon. Interconnecting strands associated with neurofilaments (arrows) are clearly observed. Some of these strands directly attach to the surface of the mitochondria (Mt). ×150,000. b, Freeze-etch replica showing the axonal cytoskeleton (AC) underlying the axolemma (AL). The characteristic meshwork of filamentous structures (arrows) is directly applied to the cytoplasmic surface of the axolemma. My: myelin sheath. ×67,000.

membranous organelles bore no morphological resemblance to the replica images of dynein ATPase in Tetrahymena cilia (*5*, *20*), suggesting that the microtubule-dynein system might not be involved in the directional translocation of membranous organelles along the microtubules inside axons.

Similar results were obtained independently by two other groups (*7, 15*). Although it is too early to precisely interpret many of the new features observed in freeze-etch replicas, this approach combined with biochemical analysis will lead to better understanding of the molecular organization of the axonal cytoskeleton.

II. AXONAL CYTOSKELETON UNDERLYING THE NON-EXCITABLE AXOLEMMA

Freeze-etch replicas of the internodal part of rat myelinated axons revealed a layer of characteristic subaxolemmal meshwork of filamentous structures about 9 nm thick, which was directly applied to the cytoplasmic surface of the axolemma (Fig. 2b). The microtubules and neurofilaments were clearly seen to be connected with this subaxolemmal meshwork through the interconnecting strands. Interestingly, this subaxolemmal meshwork resembled in appearance the cytoskeletal network underlying the human erythrocyte membrane (*16, 17*). The underlying meshwork in the erythrocyte membrane is now known to be composed mainly of the high-molecular-weight, actin-binding protein, spectrin (*Mr* 220,000 and 240,000) and actin. Recently, from immunofluorescence study, Levine and Willard have reported that a high-molecular-weight, actin-binding protein, fodrin (*Mr* 235,000 and 240,000) is localized at the periphery of the myelinated axon (*9*). It is interesting to speculate that fodrin may be the major constituent of the subaxolemmal meshwork with which actin is closely associated, analogous to the spectrin-actin complex formed in the erythrocyte membrane (*1*).

Kakiuchi and his colleagues in the course of searching for calmodulin-binding proteins in various types of cells, have isolated "calspectin" as a spectrin-like, calmodulin- and actin-binding protein from a membrane fraction of brain (*8*). Since calspectin and fodrin were considered to be the same, we have, in collaboration with Kakiuchi's group, attempted to compare the molecular shape of calspectin with that of erythrocyte spectrin using low-angle rotary-shadowing electron microscopy. Calspectin was purified from bovine brain. Calspectin solution was mixed with an equal volume of glycerol and sprayed on freshly cleaved pieces of mica (*22*). After drying *in vacuo*, the samples were rotary-shadowed with platinum-carbon at a shadowing angle of ap-

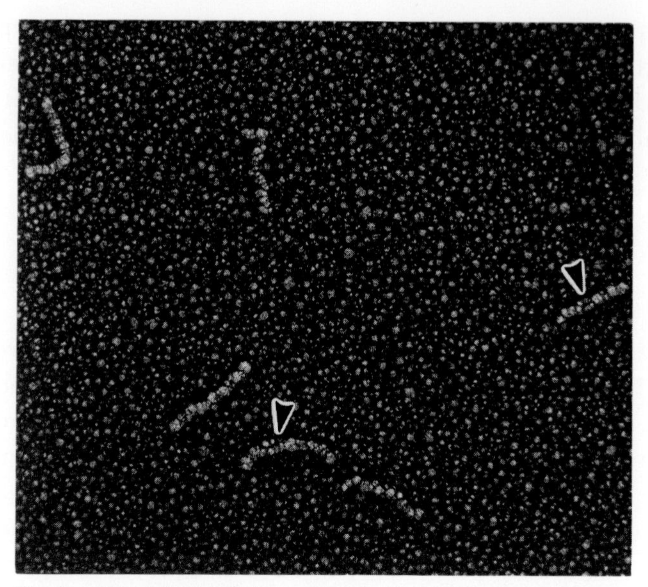

Fig. 3. Morphology of the tetrameric form of calspectin molecules in rotary-shadowed preparations. The middle portion of the tetrameric rod occasionally splits into two strands (arrowheads). ×60,000.

proximately 5°. Shadowed films were floated on water and picked up onto grids. Under rotary-shadowing electron microscopy, calspectin molecules looked like elongated rods about 220 nm in length for the tetramers and 110 nm for the dimers (*4*, *8*) (Fig. 3). Each rod showed two twisted strands, with the middle portion of the tetrameric rod more or less separated. Biochemical and electron microscopic analyses of the tetramer-dimer interconversion indicated that the tetrameric rods were formed by the head-to-head association of two dimers. This morphological appearance of the calspectin molecule was quite similar to that of erythrocyte spectrin. Furthermore, it became clear that the actin-binding sites were localized at the tail ends of calspectin molecules, similar to those of the erythrocyte spectrin (*1*). These findings suggest the functional similarity between calspectin and erythrocyte spectrin. In the erythrocyte membrane, actin exists as short filaments that appear to play a key role in formation of the continuous network of spectrin (*1*, *17*). Therefore, calspectin, with actin, may also form the cytoskeletal network underlying the non-excitable axolemma in mammalian axons, and this cytoskeletal network may connect the axonal cytoskeleton to the axolemma.

III. AXONAL CYTOSKELETON UNDERLYING THE EXCITABLE AXOLEMMA

Evidence has accumulated that the axonal cytoskeleton plays an important role in maintaining the excitability of the axolemma. The close association between the axolemma and the underlying cytoskeleton at the node of Ranvier of the mammalian myelinated axons has been well described (*13*). However, it is very difficult to biochemically analyze this association using mammalian nerves. The squid giant axon offers a highly advantageous model for correlative morphological and biochemical studies of the axonal cytoskeleton underlying the excitable axolemma. Recently, Matsumoto *et al.* have shown that a high-molecular-weight, microtubule-binding protein (*Mr* 260,000) is required to maintain the excitability of the axolemma of squid giant axons (*10*, *11*). This protein, designated here as 260 K protein, was first described by Sakai and Matsumoto as a factor for restoring the membrane excitability of squid giant axon which had been destroyed by intra-axonal perfusion of microtubule poison (*14*). Most recently, 260 K protein has been partially purified from squid nerves (*12*). It became clear that 260 K protein interacted with microtubules to make bundles *in vitro*, and that this protein did not resemble the microtubule-associated proteins (MAPs) obtained from mammalian brain. Therefore, in order to further understand the physiological roles of the 260 K protein at the molecular level, we have attempted to obtain information about its molecular shape, antigenicity, and localization in the axoplasm of squid axons.

The 260 K protein was purified from fin nerves of the squid, *Dorytheuthis bleekeri*, according to the method developed by Kobayashi. On SDS-polyacrylamide gel electrophoresis of the purified protein, a single polypeptide with a molecular weight of 260,000 daltons was shown to dominate. At first, in order to determine the localization of 260 K protein in a squid giant axon, antiserum to the purified 260 K protein was raised in rabbits. In the Ouchterlony double-diffusion plate, only the axoplasm obtained from giant axons and the purified 260 K

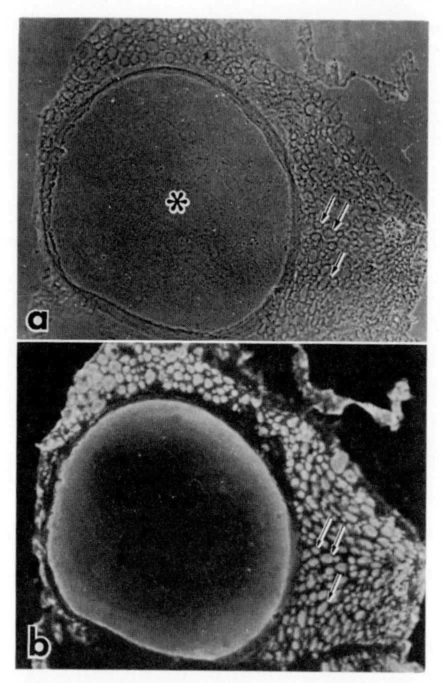

Fig. 4. Immunofluorescence microscopic localization of 260 K protein in a transverse frozen section of squid nerve containing a giant axon (*19*). a, phase-contrast microscopy. b, indirect immunofluorescence microscopy. $\times 140$. $*$: axoplasm of giant axon. arrows: small axons.

protein reacted with this antiserum with precipitin lines of antigenic identity. Then the localization of the 260 K protein was studied by indirect immunofluorescence microscopy (Fig. 4). When frozen transverse sections of squid nerves were used, intense fluorescence appeared at the peripheral axoplasm underlying the axolemma. The central region of the axoplasm in giant axons was also stained weakly. These observations indicate that 260 K protein localizes in the axoplasm underlying the excitable membrane (axolemma) and that the axoplasm in giant axons is differentiated into two regions, central and peripheral axoplasm. Judging from recent physiological studies (*10*), the peripheral axoplasm containing a large amount of 260 K protein is considered to be specialized to maintain the excitability of the axolemma. Interestingly enough, in small axons around the giant axon, the axoplasm was homogeneously stained.

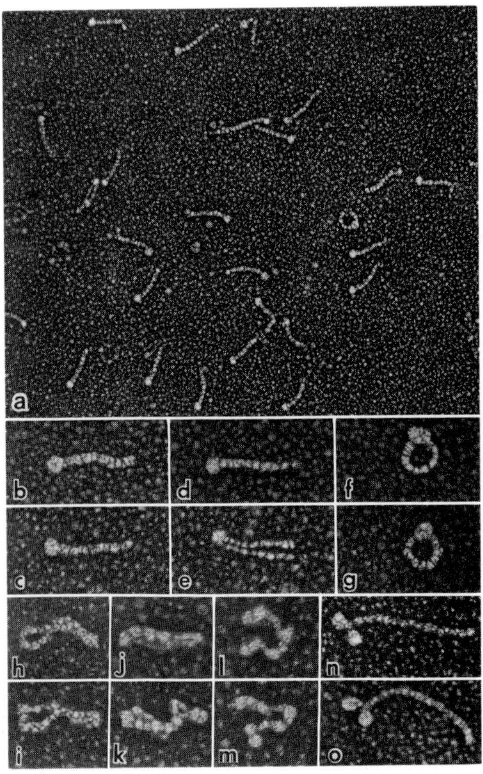

Fig. 5. Morphology of high molecular weight proteins in rotary-shadowed preparations (*19*). a–g, 260 K protein from squid nerve. 260 K protein molecules occasionally split into two strands (e) or form ring structures by head-to-tail association within one molecule (f,g). h,i, Spectrin dimers from human erythrocyte. j,k, Calspectin dimers from rat brain. l,m, Filamin dimers from chicken gizzard. n,o, Myosin from rabbit skeletal muscle. a, ×60,000. b–o, ×140,000.

Recently, various types of high molecular weight proteins have been isolated as factors for cross-linking filamentous structures to form a three-dimensional cytoskeletal network inside cells. Therefore, using low-angle rotary-shadowing technique, we compared the molecular shapes of the isolated 260 K protein, erythrocyte spectrin, brain spectrin (fodrin or calspectin), filamin, and myosin (Fig. 5). As previously described, in both erythrocyte spectrin and brain spectrin dimers, two polypeptide chains were loosely interwined and tightly joined at both ends to form a strand about 100 nm long, although the brain spectrin dimers appeared less tortuous and more rigid than erythrocyte spectrin dimers. In filamin dimers, two chains about 80 nm long appeared to be joined only at one end. Myosin molecules about 150 nm long showed a characteristic Y-shaped structure with two heads apart. In contrast to these proteins, the 260 K protein molecules appeared as a straight rod about 100 nm long with a globular head at one end. This molecule has a tendency to form a ring structure by head-to-tail association within one molecule. Furthermore, some molecules were seen to be composed of two strands which were arranged parallel with each other, suggesting that the native 260 K protein molecule might be a homodimer. These findings indicate that the 260 K protein is neither a spectrin-like, a filamin-like, nor a myosin-like protein.

The freeze-etch replica technique used in the study of the axonal cytoskeleton in mammalian myelinated axons is also thought to be potent for *in situ* morphological analysis of the peripheral axoplasm of squid giant axons. Furthermore, this approach combined with antibody staining will lead to better understanding of the molecular organization of the axoplasm underlying the excitable membrane. At present, there is very little information on how the 260 K protein regulates the membrane excitability, but we believe that further studies on the molecular organization of the peripheral axoplasm in squid giant axons will provide a clear picture of the molecular mechanism of membrane excitation.

SUMMARY

Rapid-freeze, deep-etch replicas provided three-dimensional views of the well-developed cytoskeleton inside mammalian myelinated axons, which were mainly composed of microtubules, neurofilaments, and interconnecting strands. In order to understand how the axonal cytoskeleton interacts with the axolemma, our attention has been focused on the molecular organization of the axonal cytoskeleton underlying the axolemma. At the internodal part of the mammalian myelinated axons, the specialized subaxolemmal meshwork connecting the axonal cytoskeleton to the axolemma was observed. As one possible candidate for the constituents of this meshwork, brain spectrin, calspectin or fodrin, was isolated and morphologically characterized. As a result, we were led to speculate that brain spectrin, with actin, might form the subaxolemmal meshwork, analogous to the spectrin-actin complex shown in the erythrocyte membrane. In order to study the axonal cytoskeleton underlying the excitable axolemma, the squid giant axon was used. Recently, it has been reported that a high molecular weight protein, 260 K protein, is required to maintain the excitability of the axolemma in squid giant axons. Judging from its molecular shape, it became clear that the 260 K protein was neither a spectrin-like, a filamin-like, nor a myosin-like protein. Taking this finding together with the immunofluorescence study, we concluded that the 260 K protein was a unique protein located in the axoplasm underlying the excitable membrane in squid giant axons.

Acknowledgments

We wish to thank Prof. Harunori Ishikawa, Gunma University, Prof. Shiro Kakiuchi, Osaka University, Prof. Masanori Kurokawa, University of Tokyo, Dr. Gen Matsumoto, Electrotechnical Laboratory, Dr. Jiro Usukura, University of Tokyo, Drs. Kenji Sobue and Kouichi Morimoto, Osaka University, and Dr. Takaaki Kobayashi, Jikei University, for their collaboration in this study.

REFERENCES

1 Branton, D., Cohen, C.M., and Tyler, J.M. (1981). *Cell* **24**, 24–32.
2 Ellisman, M.H. and Porter, K.R. (1980). *J. Cell Biol.* **87**, 464–479.
3 Fukuda, J., Kameyama, M., and Yamaguchi, K. (1981). *Nature* **294**, 82–85.
4 Glenney, Jr., J.R., Glenney, P., Osborn, M., and Weber, K. (1982). *Cell* **28**, 843–854.
5 Goodenough, U.W. and Heuser, J.E. (1982). *J. Cell Biol.* **95**, 798–815.
6 Heuser, J.E. and Salpeter, S.R. (1979). *J. Cell Biol.* **82**, 150–173.
7 Hirokawa, N. (1982). *J. Cell Biol.* **94**, 129–142.
8 Kakiuchi, S., Sobue, K., Kanda, K., Morimoto, K., Tsukita, S., Tsukita, S., Ishikawa, H., and Kurokawa, M. (1982). *Biomedical Res.* **3**, 400–410.
9 Levine, J. and Willard, M. (1981). *J. Cell Biol.* **90**, 631–643.
10 Matsumoto, G., Ichikawa, M., Tasaki, A., Murofushi, H., and Sakai, H. (1983). *J. Memb. Biol.* **77**, 77–91.
11 Matsumoto, G., Kobayashi, T., and Sakai, H. (1979). *J. Biochem.* **86**, 1155–1158.
12 Murofushi, H., Minami, Y., Matsumoto, G., and Sakai, H. (1983). *J. Biochem.* **93**, 639–650.
13 Peters, A. (1968). In *The Structure and Function of Nervous Tissue*, vol. 1, ed. Bourne, G.H., pp. 141–186. New York, London: Academic Press.
14 Sakai, H. and Matsumoto, G. (1978). *J. Biochem.* **83**, 1413–1422.
15 Schnapp, B.J. and Reese, T.S. (1982). *J. Cell Biol.* **94**, 667–679.
16 Tsukita, S., Tsukita, S., and Ishikawa, H. (1980). *J. Cell Biol.* **85**, 567–576.

17 Tsukita, S., Tsukita, S., Ishikawa, H., Sato, S., and Nakao, M. (1981). *J. Cell Biol.* **90**, 70–77.

18 Tsukita, S., Tsukita, S., Ishikawa, H., Kurokawa, M., Morimoto, K., Sobue, K., and Kakiuchi, S. (1983). *J. Cell Biol.* **97**, 574–578.

19 Tsukita, S., Tsukita, S., Kobayashi, T., and Matsumoto, G. (1983). *Biomedical Res.* **4**, 615–618.

20 Tsukita, S., Tsukita, S., Usukura, J., and Ishikawa, H. (1983). *J. Cell Biol.* **96**, 1480–1485.

21 Tsukita, S., Usukura, J., Tsukita, S., and Ishikawa, H. (1982). *Neuroscience* **9**, 2135–2147.

22 Tyler, J.M. and Branton, D. (1980). *J. Ultrastruct. Res.* **71**, 95–102.

23 Usukura, J. and Yamada, E. (1981). *Biomedical Res.* **2**, 177–193.

24 Usukura, J., Akahori, H., Takahashi, H., and Yamada, E. (1983). *J. Electron Microsc.* **32**, 180–185.

7

INVOLVEMENT OF PHOSPHOLIPID METABOLISM IN VISUAL TRANSDUCTION

HIROKO INOUE AND TOHRU YOSHIOKA

Department of Physiology, Yokohama City University School of Medicine, Yokohama 232, Japan

Visual receptors in vertebrates and invertebrates employ a variety of different mechanisms to translate the information accepted in rhodopsin into photoreceptor potentials. Activation of the invertebrate photoreceptor leads to a rapid and transient opening of ion channels in the plasma membrane and results in depolarization (*7*). In vertebrate photoreceptors, however, light irradiation effect on a target cell may have quite a different mechanism involving the generation of hyperpolarized potentials (*21*). These electric signals are associated with the level change in cyclic GMP (cGMP) and/or Ca^{2+} and initiate phosphorylation or conformational changes of channel proteins (*5*, *15*).

In invertebrate phototransduction systems there is no change in guanylate cyclase activity and the level of cGMP is low. Ca^{2+} may therefore play an important role (*23*) as it is generally believed that the receptors which do not influence adenylate (or guanylate) cyclase activity always provoke hydrolysis of phosphatidylinositol (PI cycle), which is thought necessary for Ca^{2+} mobilization (*6*, *14*). The phospholipid hydrolysis step, triggered by photon absorption, may thus be considered a transduction process analogous to the stimulation of adenylate cyclase that is necessary for an activated receptor to cause an increase in cAMP.

This paper describes the authors' initial attempt at genetic analysis of the phototransduction mechanism in *Drosophila*. *Drosophila* neurogenetics is a system using genetics to study the components of neural development and physiology. The technique is applicable to a broad spectrum of neural phenomena, sensory transduction, synaptic transmission and behavioral plasticity (*8*, *18*).

I. PHOSPHATIDIC ACID AND INOSITOL PHOSPHOLIPIDS

Phosphatidylinositol (PI) is a minor membrane component of the phospholipids of most animal cells (*10*). It is the precursor of phosphatidylinositol 4-phosphate (or diphosphoinositide; DPI) and phos-

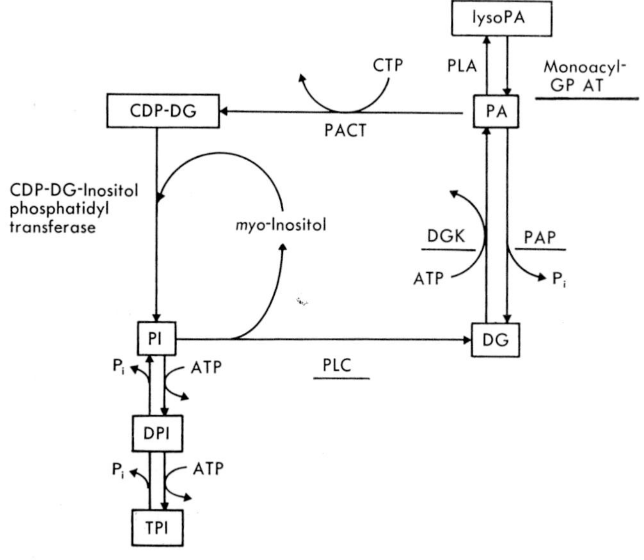

Fig. 1. Pathway of PI metabolism. Enzymes examined in the experiment on *Drosophila* compound eye are underlined. Abbreviations used are: PA: phosphatidic acid, CDP-DG: CDP-diacylglycerol, PI: phosphatidylinositol, DG: diacylglycerol or diglyceride, lysoPA: lysophosphatidic acid or monoacylglycerophosphate, DPI: phosphatidylinositol 4-phosphate or diphosphoinositide, TPI: phosphatidylinositol 4,5-bisphosphate or triphosphoinositide, PLC: phospholipase C, DGK: diglyceride kinase, PLA: phospholipase A, monoacyl-GPAT: monoacylglycerophosphate acyltransferase, PAP: phosphatidate phosphatase, PACT: phosphatidate cytidyltransferase.

phatidylinositol 4,5-bisphosphate (or triphosphoinositide; TPI), which are unique among the phospholipids because they possess one or two phosphate groups in the inositol ring. Phosphatidic acid (PA) is also a minor component, but it is the precursor of major phospholipids including PI which provide an important structural unit in the biological membrane.

The metabolic interrelationships between the three inositol phospholipids and PA are shown in Fig. 1. PA is synthesized by two pathways, one from diglyceride (DG) *via* DG kinase and another from lysoPA (or monoacylglycerophosphate) *via* monoacylglycerophosphate acyltransferase (monoacyl-GPAT). PA is hydrolyzed to DG *via* PA phosphatase. PI is formed from PA through CDP-DG and hydrolyzed to DG; this is called the PI cycle (PI turnover). DPI and TPI are formed respectively by phosphorylation of PI and DPI. The main enzymes discussed in this paper are those relating to synthesis and hydrolysis of PA and PI breakdown to DG. Enzymes examined here are those underlined in Fig. 1.

II. PI TURNOVER IN THE PHOTORECEPTOR

The breakdown and resynthesis of PI in response to a variety of well-defined extracellular stimuli have been described for a number of tissues (*9*). Synthesis of PI from [^{14}C]glycerol or [^{3}H]arachidonic acid in isolated bovine and toad retinas has been shown to exceed that of other phospholipids (*3*, *4*). *In vivo* studies have shown that the half life of PI in the retina is much shorter than that of phosphatidylcholine (PC) or phosphatidylethanolamine (PE) (*1*).

It has been shown by Anderson and Hollyfield (*2*) that the incorporation of [^{3}H]inositol into PI in retinas of *Xenopus laevis* tadpoles or young adults in short-term organ culture is stimulated by light, compared to retinas maintained under identical conditions in darkness. Over 95% of the label incorporated into lipid was in PI, and none was incorporated into retinal proteins. The ^{32}P incorporation was found by autoradiography to be localized in the outer plexiform layer, a neurophil composed primarily of horizontal cell processes (*2*).

Enhancement of synthesis and phosphorylation of PI by light in retina was reported by Schmidt (*19*, *20*). Incorporation of labeled precursors ([^{14}C]glycerol, [^{14}C]glucose or $^{32}P_i$) into PI was 2–3-fold higher in rat retinas incubated in light compared to those in dark. During brief (30 min) incubations with labeled glycerol, there was a selective increase in the radioactivity associated with PI and PA whereas, in longer (60 min) incubations, synthesis of other lipids was also enhanced in light. Phosphorylation of PI to TPI was also enhanced by light in this case and, in both light and dark, up to 40–50% of the total $^{32}P_i$ incorporation was associated with TPI.

The same kind of experiments were done by us previously in invertebrate photoreceptor systems of octopus retina (*23*). $^{32}P_i$ incorporation into DPI and PA increased with incubation time in both light and dark. The effect of light irradiation on the $^{32}P_i$ incorporation was apparently manifested in DPI and PA; the radioactivity of both bands was decreased by light exposure.

The effect of isobutylmethylxanthin (IBMX) on PI phosphorylation was studied because IBMX is known to be an inhibitor of phosphodiesterase for cyclic nucleotide and also a suppressor of the electroretinogram (ERG). The effect of 1 mM of IBMX on PI phosphorylation resulted in increased $^{32}P_i$ incorporation into PA and decreased the radioactivity of DPI. There was no effect of light irradiation on the IBMX modified PI phosphorylation. The change of intracellular level of cGMP in microvilli under light irradiation was also examined and cGMP concentration was found to be so low that no change in it was observed. The effect of cGMP on PI phosphorylation was also investigated but the phosphorylation was not appreciably affected.

The suppressive effect of IBMX on ERG production was explained as a regulation of Ca binding ability of the membrane, which was expressed as

$$\varphi = \frac{\mathrm{DPI}+\mathrm{TPI}}{\mathrm{PA}+\mathrm{PI}+\mathrm{DPI}+\mathrm{TPI}}$$

The value of φ of dark adapted retina membrane was 0.84 and that of IBMX-treated was 0.55 (*23*).

III. VISUAL MUTANT OF *DROSOPHILA MELANOGASTER*

One approach to revealing the complex system involved in the signal

TABLE I
Some Properties of Three Alleles of *norpA* Mutant of *Drosophila melanogaster*

	Normal		*norpA*	
		JM11	*SB37*	*EE5*
E R G				
Phototaxis	+++	+	±	−
Rhabdomere size (μm^2)	32±6	>	>	11±6
Number of intramembranous particles ($/\mu m^2$)	3000	>	>	480

All data are from the laboratory of Dr. Yoshiki Hotta of the University of Tokyo. Some have been published in the reference papers *10* and *11*.

transduction is to alter the elements by specific mutation. *Drosophila* mutant with abnormal visual systems has been used to study the neurogenetic aspect of visual transduction (*11*). A *Drosophila* mutant, *norpA* (no receptor potential A), has been isolated which does not produce receptor potentials (*17*). Its rhabdomere is formed and contains rhodopsin molecules (*16*); therefore, it is believed that the *norpA* gene is essential for the transduction process between photon absorption and membrane ion permeability changes. Freeze fracture electron microscopy (*10*) revealed that the *norpA* mutant rhabdomeres have a reduced amount of rhabdome specific proteins (*11*) and fewer than normal intramembranous particles (Table I). As shown, the Canton-S strain of *Drosophila melanogaster* was used as a standard from which mutation was induced with an alkylating agent, ethyl methansulfonate. The three alleles of *norpA* gene used were $norpA^{EE5}$ (*EE5*), $norpA^{SB37}$ (*SB37*) and $norpA^{JM11}$ (*JM11*). Mutants called $rdgA^{KO14}$ (receptor degeneration A) and *sine oculis* (without compound eye) were also examined to confirm whether or not the abnormality observed in *norpA* mutant is localized on the retinular cell.

IV. LOCALIZATION OF PA IN THE RETINULAR CELL MEMBRANE OF *DROSOPHILA*

The aim of biochemical research into the mode of action of the photoreceptor as well as the neurotransmitter receptor, is to determine the sequence of events between photon absorption and the final electrical response of the cell. It is generally expected that localization of the special molecules in the cell suggest the cell's special function; for example, rhodopsin is localized on the visual cell and is known to play an important role in photoreception.

Localization of PA in the retinular cell membrane of *Drosophila* was found by following two different kinds of analysis.

(i) Genetic analysis (*24*): In a phosphorylation experiment using fly head homogenates, it was found that $^{32}P_i$ of [γ-^{32}P]ATP was incorporated into PA of the retinular cell membrane selectively, because PA was not labeled when head homogenate of *rdgA*^{KO14} was employed (Fig. 2b). This mutation has been shown to cause post-eclosional degeneration of peripheral retinular cells and their rhabdomeres without affecting the morphology of other cells. Similar results were also obtained by examining the *sine oculis* mutant, which entirely lacks compound eyes and also a part of the associated optic ganglia (Fig. 2c).

(ii) Manual dissection of freeze-dried *Drosophila* eye: According to

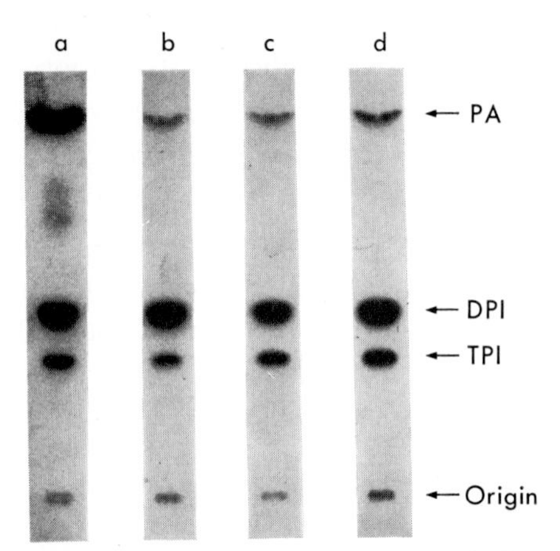

Fig. 2. Autoradiograph after TLC of phosphorylated phospholipid. Phosphorylation was started by adding [γ-^{32}P]ATP (10 μCi/ml) to the head homogenate (200 μg protein/ml) and stopped by adding an equal volume of ice-cold perchloric acid to the reaction medium. Labeled phospholipid was extracted by acidic chloroform/methanol solution and developed with a mixture of chloroform/acetone/methanol/acetic acid/water (40:17:15:12:8, v/v). a: normal strain (CS), b: *rdgA*^{KO14} (mutant without peripheral rhabdomere), c: *sine oculis* (mutant without eye), d: *norpA*^{EE5} (mutant without receptor potential).

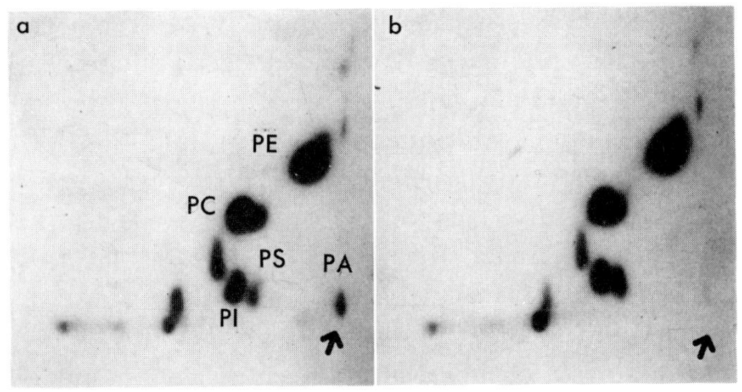

Fig. 3. Autoradiograph of two-dimensional TLC pattern of compound eye from dried head (a) and brain (b). Labeling of phospholipids was carried out by introduction of $^{32}P_i$ into *Drosophila* by ingestion. Labeled flies were frozen in acetone-dry ice solution and evacuated for 2 days. Compound eye and brain were dissected mechanically by fine needle. Labeled phospholipid was extracted by acidic chloroform/methanol solution and developed in the first dimension with chloroform/methanol/28% NH_4OH (65:35:5, v/v) and in the second dimension with chloroform/acetone/methanol/acetic acid/water (50:20:10:10:5).

previous data, it is plausible that PA is localized in the retinular cell of the *Drosophila* compound eye. In order to obtain clear evidence of this, all kinds of phospholipids were labeled by introduction of $^{32}P_i$ into the fly body by ingestion. The labeled flies were frozen in an acetone-dry ice solution and were evacuated in a vacuum for 2 days. A fully dried head was dissected into two parts, the compound eye and the brain. As is shown in Fig. 3, PA is localized only in the compound eye, not in the brain (*25*).

These results led us to the conclusion that PA in the retinular cell is localized in the photoreceptor membrane. If a molecule is localized in a particular cell, a determining factor of the cell function must be ascribed to that molecule. This anticipation was successfully realized in the photoreceptor function of *Drosophila*.

V. PA AND PHOTORECEPTOR POTENTIALS

We obtained evidence supporting the notion that PA is necessary for receptor potential production. The argument for this view is presented here in detail. The ERG of the three alleles of *norpA* were compared with a wild type strain. ERG of normal strain is shown in the upper column in Table I for a short-term flash; an early corneal negative dip is followed by a sharp positive peak, then a longer lasting negative wave. The early negative dip and late negative component are roughly proportional in size at various flash intensities. Short flash ERG of the *norpA* mutant is quite abnormal; in *JM11* and *SB37* the early negative dip disappeared and a negative component barely survived. The ERG of *EE5* was completely abolished. Relative ERG sizes are 1:0.5:0.2:0 for CS:*JM11*:*SB37*:*EE5*.

The PA content of these alleles of *norpA* mutant obtained by *in vitro* phosphorylation experiment, chemical analysis and *in vivo* phosphorylation are comparable. The phosphorylation experiment *in vivo* resulted in a decrease in PA radioactivity in the mutant: 1(CS):0.50(*JM11*):0.53(*SB37*):0.13(*EE5*) (*24*). Chemical analysis of PA in 200 heads of *Drosophila* did not give us quantitative data because the PA content was below the detectability level. The PA spot of CS head homogenate, however, was detected by iodine vapor although others were not. Final determination was obtained by counting the radioactivity of PA labeled by the ingestion of radioactive P_i. Percent radioactivity of labeled PA

TABLE II
Comparison of Enzymatic Activity Relating to PA Synthesis Between Normal Strain and Mutant

	CS	*norpA*^{JM11}	*norpA*^{EE5}
DG kinase[a)]	100	70±8	12±3
PA phosphatase[b)]	100	98±9	114±8
Monoacyl-GPAT[c)]	100	115±3	97±6

* The value of enzymatic activity for normal (CS) is normalized to 100 and data for the mutants are expressed as mean % of the ratio ±S.D. %.
a) DG kinase activity was determined by measuring incorporation of $^{32}P_i$ from [γ-^{32}P]ATP into PA. The basic reaction medium contained 100 mM KCl, 10 mM NaCl, 10 mM $MgCl_2$, 10 mM KF, 1 mM EGTA, 2 mM DG, 1 mM ATP and 20 mM phosphate buffer (pH 6.8). Twenty fly heads were homogenized in 0.2 ml of the basic reaction medium. Phospholipid was extracted and analyzed on TLC. PA spot was scraped and the radioactivity was counted.
b) PA phosphatase activity was determined by measuring the amount of phosphorus liberated from PA. Head homogenate (0.4 mg protein) was incubated with PA (4 mM) containing Tris buffer (pH 7.4) solution (final volume, 0.2 ml) for 30 min at room temperature. c) Monoacyl-GPAT was assayed by measuring radioactive PA which was synthesized *in vitro* from monoacyl-GP. Reaction medium (0.2 ml) contained 100 mM Tris-HCl buffer (pH 7.4), 0.1 mM monoacyl-GP, [1-^{14}C]oleoyl-Coenzyme A and 0.17 mg protein of fly head homogenate. Reaction was performed for 2 min at room temperature. Analysis and radioactivity measurement of PA was as described in (a).

in all phospholipids of the normal fly head was about 0.53% and that of *EE5* a mere 0.12%. Thus PA content exclusively correlates with the photoreceptor potential.

VI. AN ACTIVE PATHWAY OF PA FORMATION IN *DROSOPHILA* COMPOUND EYE

PA is synthesized either from DG *via* DG kinase or from lyso PA *via* monoacyl-GPAT and is hydrolyzed by PA phosphatase. The activities of DG kinase, monoacyl-GPAT and PA phosphatase were examined under defined conditions between normal strain and the *norpA* mutant. Results are listed in Table II and the following conclusions are derived from the data: (1) PA is mainly synthesized from DG via DG kinase, (2) DG kinase is localized in the retinular cell of the *Drosophila* and (3) severity of the defect in DG kinase is in parallel with the deficit in the receptor potential (*25*).

VII. PI CYCLE IN THE *DROSOPHILA* COMPOUND EYE OF *norpA*

Many researchers consider that the PI cycle begins with the hydrolysis of the lipid by phospholipase C (PLC). The PLC activity of three alleles of *norpA* was measured and activity was found to be entirely absent in the mutant (*25*). Generally, PLC hydrolyzing PI requires Ca ions and the enzyme is soluble, but the membrane bound form of the enzyme is found only in rat liver and brain (*12, 13*). In the *Drosophila* compound eye, PLC activity was found exclusively in the microsomal fraction precipitated at $100{,}000 \times g$ for 60 min. Therefore, *norpA* has no receptor potential, low enzymatic activity in DG kinase and no activity in PLC. These results led us to the conclusion that there is no PI cycle in the mutant of *Drosophila, norpA*. An understanding of the role of the PI cycle in the production of photoreceptor potential requires further knowledge of the enzymatic deficiency in various visual transduction mutants of *Drosophila*.

SUMMARY

A visually defective mutant of *Drosophila, norpA* (without receptor potential) has been shown to lack photoreceptor potentials, although the photoreceptor membrane contains rhodopsin and has normal morphology. Lipid analysis of the retinular cell membrane showed a complete lack of phosphatidic acid (PA) in *norpA*. Investigation of the pathway for PA formation revealed that PA was synthesized from diglyceride (DG) via DG kinase. The DG kinase activity was markedly reduced in the mutant and the abnormality of the enzyme was closely related to the severity of the mutant syndrome. Defects in phospholipase C activity were revealed in the retinular cell of the mutant. The phospholipid metabolism is thus shown to be unsound in the *norpA* mutant.

Acknowledgments

This study was supported in part by a Grant-in-Aid from the Ministry of Education, Science and Culture of Japan and by the Nissan Science Foundation (No. 811004). The authors thank Dr. Yoshiki Hotta of the University of Tokyo who collaborated in the *Drosophila* work.

REFERENCES

1 Anderson, R.E., Mande, M.B., and Kelleher, P.A. (1980). *Biochim. Biophys. Acta* **620**, 236–246.
2 Anderson, R.E. and Hollyfield, J.G. (1981). *Biochim. Biophys. Acta* **665**, 619–622.
3 Bazan, H.E.P. and Bazan, N.G. (1975). *Life Sci.* **17**, 1671–1678.
4 Bazan, H.E.P. and Bazan, N.G. (1976). *J. Neurochem.* **27**, 1051–1057.
5 Bownds, M.D. (1980). *Photochem. Photobiol.* **32**, 487–490.
6 Downes, C.P. (1983). *Trends Neurosci.* **6**, 313–316.
7 Fuortes, M.G.F. and O'Bryan, P.M. (1972). In *Handbook of Sensory Physiology*, ed. Fuortes, M.G.F., vol. VII/2, pp. 279–319. Berlin: Springer-Verlag.
8 Hall, J.C. (1982). *Q. Rev. Biophys.* **15**, 223–479.
9 Hawthorne, J.N. (1982). In *Phospholipids*, ed. Hawthorne, J.N. and Ansell, G.B., pp. 263–278. Amsterdam: Elsevier.
10 Hirosawa, K. and Hotta, Y. (1982). In *Structure of the Eye*, ed. Hoolyfield, J.G., pp. 45–53. New York: Elsevier Biomedical.
11 Hotta, Y. (1979). In *Mechanism of Cell Change*, ed. Ebert, J.D. and Okada, T.S., pp. 169–181, New York: JohnWiley & Sons, Inc.
12 Irvine, R.F., Hemington, N., and Dawson, R.M.C. (1978). *Biochem. J.* **122**, 605–607.
13 Matsuzawa, Y. and Hostetler, K.Y. (1980). *J. Biol. Chem.* **255**, 646–652.
14 Michell, R.H. (1975). *Biochim. Biophys. Acta* **415**, 81–147.
15 O'Brien, D.F. (1982). *Science* **218**, 961–966.
16 Ostroy, S.E. (1978). *J. Gen. Physiol.* **72**, 717–732.

17 Pak, W.L. (1975). In *Handbook of Genetics*, ed. King, R.C., vol. 3, pp. 703–733. New York: Plenum Press.

18 Pak, W.L. (1979). In *Neurogenetics: Genetic Approach to the Nervous System*, ed. Breakefield, X.O., pp. 67–99. New York, North Holland: Elsevier.

19 Schmidt, S.Y. (1983). *J. Neurochem.* **40**, 1630–1638.

20 Schmidt, S.Y. (1983). *J. Biol. Chem.* **258**, 6863–6868.

21 Tomita, T. (1972). In *Handbook of Sensory Physiology*, ed. Fuortes, M.G.F., vol. VII/2, pp. 483–511. Berlin: Springer-Verlag.

22 White, D.A. (1973). In *Form and Function of Phospholipid*, ed. Ansell, G.B., Dawson, R.M.C., and Hawthorne, J.N., pp. 441–482, Amsterdam: Elsevier.

23 Yoshioka, T., Inoue, H., Takagi, M., Hayashi, F., and Amakawa, D. (1983). *Biochim. Biophys. Acta* **755**, 50–55.

24 Yoshioka, T., Inoue, H., and Hotta, Y. (1983). *Biochem. Biophys. Res. Commun.* **111**, 567–573.

25 Yoshioka, T., Inoue, H., and Hotta, Y. (1983). *J. Neurochem.* **41**, S120.

8

PHOSPHOINOSITIDES IN THE AUDITORY SYSTEM

JOCHEN SCHACHT

Kresge Hearing Research Institute, University of Michigan, Michigan 48109, U.S.A.

Phosphorylation reactions are a ubiquitous mechanism of controlling physiological activities of cells. This type of regulation may affect the processing of genetic information (*via* phosphorylation of histones), enzymatic activity, and properties of membranes. In most instances the substrates of phosphorylation have been identified as proteins. In addition, there is growing evidence that phosphorylation/dephosphorylation cycles of phospholipids play a role in information processing at the membrane.

Phospholipids provide the structural framework of biological membranes and appear to be important regulators of membrane permeability (*5*, *10*). Polyphosphoinositides, phosphatidylinositol 4-phosphate and phosphatidylinositol 4,5-bisphosphate, lipids derived from phosphatidylinositol by two sequential kinase reactions, have long been speculated to control functions of the plasma membrane *via* regulation of calcium-binding. While a precise role for these lipids remains to be established, current evidence links alterations of their metabolism to hormone-receptor interactions (*3*, *11*). Breakdown of polyphosphoinositides in the plasma membrane may be a primary receptor-coupled event changing membrane conformation and possibly calcium levels. Resynthesis of phosphoinositides from the resulting diglyceride would complete the cycle. In addition, interconversions of phosphoinositides are possible by phosphomonoesteratic cleavage, phosphatidylinositol 4,5-bisphosphate ⟶ phosphatidylinositol 4-phosphate ⟶ phosphatidylinositol followed by the corresponding kinase reactions (refer also to the chapters by Imai, Inoue, Irvine, and Sokabe in this volume).

Our interest in the function of polyphosphoinositides in the processing of acoustic stimulation stems from the apparent involvement of these lipids in aminoglycoside-induced hearing loss. The aminoglycoside-aminocyclitol antibiotics (neomycin, gentamicin, and related compounds) possess a rather specific toxicity for the inner ear and the kidney. Based on our work over the past years (see *6*, *12*, and *13* for reviews) we have formulated the hypothesis that polyphosphoinositides are specific binding sites for aminoglycosides. The configuration of the negative charges of phosphatidylinositol bisphosphate allows a specific three-point binding to the positively charged aminoglycoside and hydrogen bonding may further stabilize the complex. Our hypothesis suggests that the first step of aminoglycoside toxicity is a competitive and reversible displacement of calcium from various binding sites. This step which inhibits calcium-dependent membrane functions is followed by a specific complex formation with phosphatidylinositol bisphosphate. This will disrupt membrane structure and alter membrane permeability.

If polyphosphoinositides mediate aminoglycoside ototoxicity then one might speculate that these lipids play a crucial role in auditory function. Such a role would be expected at the level of the receptor cells. The binding affinity of aminoglycosides to polyphosphoinositides correlates with aminoglycoside-induced suppression of the cochlear microphonic, a potential which is thought to originate in the sensory cells (*9*). We thus tested whether acoustic stimulation would alter phosphoinositide metabolism in auditory receptors.

I. PHOSPHOINOSITIDES IN THE GUINEA PIG COCHLEA

In the first series of experiments we studied phospholipid labelling in the inner ear of the guinea pig. The precursor, [^{32}P] orthophosphate ($^{32}P_i$) was introduced into the cochlea by perfusion of the fluid spaces

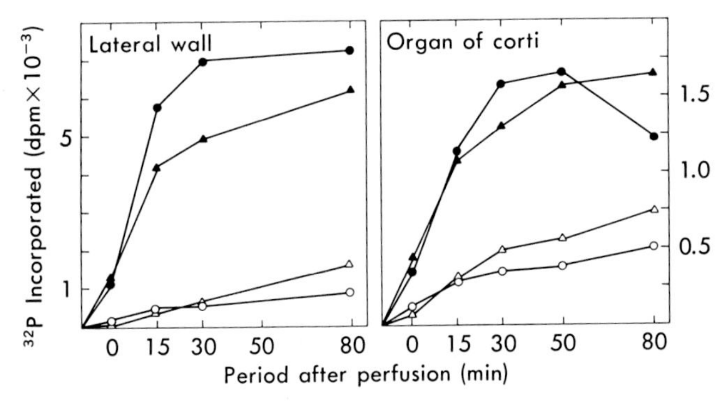

Fig. 1. Phospholipid labeling in the guinea pig cochlea. Guinea pigs received a perilymphatic perfusion with 250 μC (0.1 mM) $^{32}P_i$ in artificial perilymph for 10 min. Flow was then stopped and $^{32}P_i$ remained in the cochlea for the indicated times ("Period after Perfusion"). ○—○, phosphatidate; △—△, phosphatidylinositol; ▲—▲, phosphatidylinositol 4-phosphate; ●—●, phosphatidylinositol 4,5-bisphosphate.

with "artificial perilymph". This technique yielded high levels of radiotracer in the inner ear while preserving its function which was monitored by recording the cochlear microphonic potential (*16*, *17*).

Polyphosphoinositides were rapidly and highly labeled (Fig. 1). The tissues of the "lateral wall" consist of spiral ligament and stria vascularis which is generally considered to be involved in the secretion and maintenance of the endolymph fluid. The "organ of Corti" contains the inner and outer hair cells—the sensory cells—as well as numerous supporting elements. Lipid labeling is similar in both structures and essentially conforms to the pattern generally observed in secretory and sensory tissues: $^{32}P_i$ is predominantly incorporated into the polyphosphoinositides while phosphatidate and phosphatidylinositol are labeled less. Furthermore, as demonstrated for other nervous or secretory tissues, it seems that turnover of the monoesterified phosphate groups of the polyphosphoinositides *via* monoesterase and kinase reactions is primarily responsible for the rapid ^{32}P-incorporation. A contribution by other pathways seems negligible as both the incorporation of radio-labeled inositol (*2*) or glycerol (*16*) into these lipids proceeds at a much slower rate and in the reverse order, phosphatidylinositol ≫ phosphatidylinositol phosphate > phosphatidylinositol bisphosphate.

Exposure of the perfused ear to moderate (ambient) or loud (100 dB) sound did not change the pattern of ^{32}P-incorporation (Table I).

Neither the ratio of phosphatidylinositol bisphosphate to phosphatidylinositol phosphate nor to the other lipids was affected. An interpretation of these results was, however, hampered by the complexity of the mammalian inner ear. If only lipids of hair cells respond to stimulation then the lipid content of supporting cells and other structures in the preparation could have masked such a localized effect.

With this in mind, we explored polyphosphoinositide metabolism in less complex hair cell systems, the lateral line organ of fish and the ear of the Noctuid moth.

II. PHOSPHOINOSITIDES IN THE LATERAL LINE ORGAN

The lateral line organ and the mammalian labyrinth have evolved from a common octavo-lateralis system. While simpler in structure, the lateral line is similar to the labyrinth in morphology and receptor mechanisms and can be considered a valid model for studies of auditory transduction. Furthermore, the lateral line canal organ of fish is easily accessible to experimental manipulations and controlled stimulation (*4*). In a series of experiments in the laboratory of Dr. Å. Flock we explored phospholipid labeling in this model.

Intact and responsive sensory epithelia from the burbot (*Lota*

TABLE I
Effect of Sound on Phospholipid Labeling in the Guinea Pig Cochlea

Period after perfusion (min)	Quiet	Ambient	100 dB
		$PhIP_2/PhIP$	
10	1.2±0.3 (5)	1.2±0.2 (6)	1.1±0.1 (5)
30	1.4±0.5 (6)	1.2±0.2 (6)	—
60	1.3±0.3 (10)	—	1.4±0.2 (7)
		$PhIP_2/PhI+PhA$	
10	2.4±0.5 (4)	2.2±0.2 (5)	2.2±0.4 (5)
30	2.4±0.8 (4)	2.5±0.8 (5)	—
60	2.2±1.2 (10)	—	2.2±0.4 (7)

Conditions as in Fig. 1. Sound exposure was given during the "Period after Perfusion" through a sound tube placed in the ear canal. "Ambient" refers to laboratory environment of approx. 50–60 dB. For the "Quiet" condition the incudostapedial joint was severed. Numbers are means±S.D. with number of experiments in parentheses (From Stockhorst and Schacht, unpublished). $PhIP_2$, phosphatidylinositol 4,5-bisphosphate; PhIP, phosphatidylinositol 4-phosphate; PhI, phosphatidylinositol; PhA, phosphatidate.

TABLE II
Effect of Stimulation on Phospholipid Labeling in the Lateral Line Organ

Lipid	Control	Stimulated
	(% of total lipids)	
$PhIP_2$	39 ± 7	37 ± 11
PhIP	35 ± 4	39 ± 4
PhA+PhI	22 ± 5	23 ± 8
PhC	4 ± 3	3 ± 2
PhE	<1	<1

Lateral line organs of *Lota vulgaris* were incubated for 60 min at 22–24°C with 0.5 mCi $^{32}P_i$/ml medium (115 mM NaCl, 3.4 mM KCl, 1.6 mM $CaCl_2$, 2.5 mM $MgCl_2$, 0.1 mM NaH_2PO_4, 30 mM glucose, 20 mM Tris acetate, final pH 7.4). Stimulus was 90 Hz, .1 sec on/.1 sec off (*4*). Values are means±S.D. of 7 (control) and 5 (stimulated) experiments, respectively. The study was carried out in collaboration with Dr. Å. Flock, Stockholm. Abbreviations: see Table I; PhC, phosphatidylcholine; PE, phosphatidylethanolamine.

vulgaris) were incubated with $^{32}P_i$ with or without stimulation. Subsequently the sensory epithelium was dissected free of surrounding non-sensory tissue and nerve fibers and analyzed. Lipid labeling was similar to that found in the guinea pig ear (Table I), namely, high ^{32}P-incorporation into polyphosphoinositides (~75% of total ^{32}P-lipid). In contrast to the guinea pig, a low but distinct incorporation into phosphatidyl choline and ethanolamine was seen. The surrounding non-sensory tissues demonstrated a similar labeling pattern.

As a stimulus, vibration was applied to the preparation (*4*) while the microphonic output was monitored (typically 20–30 μV). There was no effect on the pattern of lipid labeling (Table II), another finding consistent with the results in the guinea pig cochlea.

III. PHOSPHOINOSITIDES IN THE EAR OF THE NOCTUID MOTH

The acoustic receptors in the ear of the moth are primary sensory cells with simple morphological and electrophysiological features and thus well suited for the investigation of biochemical correlates of sound stimulation (*7*). Particular advantages are a favorable ratio of sensory cells to supporting structures; no efferent or afferent synapses; receptor and generator potential that are identical and which lead directly to the production of action potentials. In this auditory structure polyphosphoinositides were the predominantly labeled lipids (Fig. 2) with

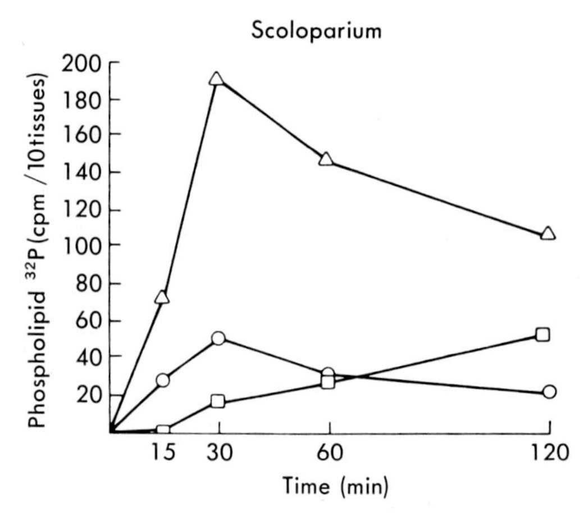

Fig. 2. Phospholipid labeling in the moth ear. One hundred μCi of $^{32}P_i$ (carrier-free, in a volume of 7 μl) were injected into the sub-epimeral air space as previously described (*8*). Moths were killed at the times indicated by an injection of glutaraldehyde followed by microwave irradiation. Ten tissues were pooled for each point. △—△, phosphatidylinositol 4,5-bisphosphate; ○—○, phosphatidylinositol 4-phosphate; □—□, phosphatidic acid and phosphatidylinositol.

a remarkably high ratio of phosphatidylinositol bisphosphate to phosphatidylinositol phosphate.

The ear of the moth is most sensitive to frequencies in the ultrasonic range (40–45 kHz). Especially interesting is the fact that pulsed tones trigger both generator and action potentials, while the presentation of a continuous tone rapidly leads to adaptation of spike activity. This leaves the receptor potential as the only bioelectric event making the organ an ideal system for its study. A continuous tone stimulus did not alter ^{32}P-labelling which was consistent with the findings in the guinea pig cochlea and the burbot lateral line. This strongly suggested that activation of the receptor cell does not lead to increased polyphosphoinositide turnover.

We did, however, pursue the moth ear further as a model using pulsed tones which also trigger and maintain the action potential. When a pulsed tone of 74 dB was given, lipid labelling was specifically altered (Table III). Radioactivity in ATP did not differ between sound-exposed animals and matched controls and neither did ^{32}P-incorporation into phosphatidylinositol plus phosphatidate. There was, however, a specific elevation of polyphosphoinositide labeling of approximately

TABLE III
Effect of Sound on ^{32}P-Incorporation in the Ear of the Noctuid Moth

Labeled compound	Stimulation by sound (% of control)
ATP	105±7
$PhIP_2$	143±18
PhIP	142±22
PhI, PhA	107±19

Moths received 175 μCi of $^{32}P_i$ (carrier free, Volume=7 μl) and were killed after 30 min. Stimulation during this time was a pulsed tone signal (40 kHz, 74 dB SPL, 50 msec on, 1 cycle/sec). Numbers are means±S.D. of 5 independent experiments; per experiment, a pool of 10–12 tissues (scoloparium) was analyzed. Significance of difference between control and stimulated groups by χ^2 ranking: $p<0.035$ for polyphosphoinositides (*8*). Abbreviations: see Table I.

40% for each lipid. Furthermore, the effect was confined to the sensory structure since polyphosphoinositides in an adjacent non-sensory tissue (nodular sclerite) remained unaffected. Thus the changes in polyphosphoinositide labeling appeared to correlate with spike activity rather than the generator potential (*8*). The preparations from the guinea pig cochlea and the burbot lateral line had been dissected to minimize the content of nerve fibers and thus could not have detected a stimulated labeling associated with action potentials.

IV. CONCLUSION

Polyphosphoinositides are present in auditory receptors of mammals, fish and insects. Their turnover in these structures, as measured by ^{32}P-labeling is more rapid than that of any other lipid, and probably involves phosphorylation cycles of the monoesterified phosphate groups *via* phosphomonoesterase and kinase reactions.

Stimulation of these receptors does not alter lipid labeling in relation to receptor (or generator) potential activity in the sensory cells. While some caution in the interpretation of the results from guinea pig and to a lesser extent the lateral line might be indicated due to the complexity of the organs, the experiments in the simple auditory receptor of the Noctuid moth appear conclusive. In addition, the data from the moth ear are also consistent with the hypothesis that polyphosphoinositides play a role in events related to neural excitation.

Recently it was reported that light stimulates turnover of phosphoinositides in the retina (*14*) by two pathways: synthesis and turnover of phosphatidylinositol, and a phosphorylation cycle involving phosphatidylinositol bisphosphate. The latter appeared not to be associated with the photoreceptor cells proper (*15*) but rather with horizontal cells (*1*). It is intriguing that also in the visual system a stimulated turnover of polyphosphoinositides can be observed which is not correlated with information processing at the receptor cells.

Our results should not be interpreted to restrict the possible role of polyphosphoinositides in auditory structures to an involvement in action potentials. If the initial metabolic event is a diesteratic breakdown of the phosphoinositides it is conceivable that the technique of tracing their synthesis by ^{32}P-incorporation is inadequate. Alternatively, an altered turnover may not be associated with the function of these lipids or it may be triggered by other physiological stimuli. Thus, it remains unresolved why the potency of aminoglycosides to suppress sensory cell function is so well correlated with their affinity to polyphosphoinositides (*9*) which initially suggested an involvement of the lipids in sensory cell physiology. Furthermore, recent studies on cochlear lipid metabolism demonstrated that noise exposure of the guinea pig prior to *in vitro* labelling of lipids resulted in enhanced ^{32}P-incorporation into phosphatidylinositol bisphosphate and phosphatidate (*18*). While this is an intriguing finding, it remains to be established how it relates to our observations. More information is needed to integrate all evidence and formulate a rational hypothesis of the function of these lipids in the auditory system.

SUMMARY

Polyphosphoinositides are rapidly labeled lipids of hair cell systems of mammals, fish, and insects. Indirect evidence from ototoxicity studies suggested an involvement of these lipids in the function of sensory cells. Tests of this hypothesis led to the conclusion that receptor (generator) potential activity in sensory cells is not associated with increased polyphosphoinositide turnover. The results are, however, compatible with an involvement of polyphosphoinositide turnover in action potential activity.

Acknowledgments

The author's research is supported by NIH research grant NS-13792 and program project grant NS-05785. Special thanks to Ms. Sue Pierson for her help in preparing this manuscript.

REFERENCES

1 Anderson, R.E., Maude, M.B., Kelleher, P.A., Rayborn, M.E., and Hollyfield, J.G. (1983). *J. Neurochem.* **41**, 764–771.
2 Anniko, M. and Schacht, J. (1981). *Int. J. Biochem.* **13**, 951–953.
3 Farese, R.V. (1983). *Endocr. Rev.* **4**, 78–95.
4 Flock, Å. (1965). *Acta Otolaryngol.* Suppl. **199**, 1–90.
5 Green, D.E., Fry, M., and Blondin, G.A. (1980). *Proc. Natl. Acad. Sci. U.S.* **77**, 257–261.
6 Humes, H.D., Weiner, N.D., and Schacht, J. (1982). In *Nephrotoxicity and Ototoxicity of Drugs*, ed. Fillastre, J.P., pp. 333–343. Paris: INSERM.
7 Kilian, P. and Schacht, J. (1977). In *Inner Ear Biology*, ed. Portmann, M. and Aran, J.M., pp. 167–168. Paris: INSERM.
8 Kilian, P. and Schacht, J. (1980). *J. Neurochem.* **34**, 709–712.
9 Lodhi, S., Weiner, N.D., Mechigian, I., and Schacht, J. (1980). *Biochem. Pharmacol.* **29**, 597–601.
10 Michell, R.H. (1975). *Biochim. Biophys. Acta* **415**, 81–149.
11 Michell, R.H., Jafferji, S.B., and Jones, L.M. (1977). *Adv. Exp. Med. Biol.* **85**, 447–464.
12 Schacht, J., Lodhi, S., and Weiner, N.D. (1977). In *Membrane Toxicity*, ed. Miller, M.W. and Shamoo, A.E., pp. 191–208, New York: Plenum Press.
13 Schacht, J., Wang, B., and Weiner, N.D. (1983). *Proc. 13th Int. Congr. Chemother.* **105**, 39–44.
14 Schmidt, S.Y. (1983). *J. Neurochem.* **40**, 1630–1638.
15 Schmidt, S.Y. (1983). *J. Biol. Chem.* **258**, 6863–6868.
16 Tachibana, M., Anniko, M., and Schacht, J. (1983). *Acta Otolaryngol.* **96**, 31–38.
17 Takada, A. and Schacht, J. (1982). *Hearing Res.* **8**, 179–186.
18 Yanagisawa, K., Yoshioka, T., Inoue, H., Hayashi, F., and Tanaka, Y. (1982). *Neurosci. Lett.* **9**, S 113.

Acknowledgments

The author's research is supported by NIH research grant NS-[illegible] and program project grant NS-[illegible]. [illegible]

[illegible]

REFERENCES

[illegible]

9

CHARACTERIZATION OF BIOMEMBRANES WITHIN THE COCHLEA

MASAYOSHI TACHIBANA

Department of Otolaryngology, Kyoto Prefectural University of Medicine, Kyoto 606, Japan

Evolutional inertia has provided two functionally distinct tissues within the mammalian cochlea, the organ of Corti and the vascular stria. The cochlea is a coiled tube divided into three fluid-filled compartments: scala vestibuli, scala tympani and scala media. The former two contain perilymph of high sodium and low potassium concentration while the last one, the central triangular compartment bounded by Reissner's membrane superiorly, the basilar membrane inferiorly and the vascular stria laterally, contains endolymph of high potassium and low sodium concentration and the organ of Corti. The organ of Corti rests on the basilar membrane and consists mainly of three rows of outer hair cells, one row of inner hair cells, supporting cells (Deiters, Hensen and Claudius) and pillar cells. Surrounding interstitial fluid, called corti-lymph, is continuous through the basilar membrane with perilymph of the scala tympani. The outer and inner hair cells are mainly innervated by efferent and afferent nerves, respectively. The superior surface of the organ of Corti, the reticular lamina, separates the organ from endo-lymph of the subtectorial space, except for stereocilia (hairs) of hair cells. The vascular stria rests on the spiral ligament and consists mainly of three kinds of epithelial cells (marginal, intermediate and basal) in close association with intraepithelial capillaries. The vascular stria is thought to secrete endolymph into the "endolymphatic" space of scala media.

It is generally accepted that the endocochlear potential (EP, +80 mV) is the sum of a positive element (+120 mV) generated as potassium secretion potential in the vascular stria and a negative element (−40 mV) as potential associated with the diffusion of potassium from endolymphatic to perilymphatic space. According to the "battery theory" proposed by Davies (*1*) and modified by others, EP (+80 mV) adds to the dc resting potential in the hair cells (−60 mV) in a series circuit to produce a 140 mV voltage source that generates current through the hair cells. The current flow is thought to be modulated by a variable resistor in the stereocilia or hair cell proper, which changes its resistance with the act of shearing. Thus the organ of Corti and the vascular stria play distinctly different functional roles; the tissues also differ in their susceptibility to ototoxic drugs. For example, aminoglycoside antibiotics such as neomycin, kanamycin and gentamicin affect primarily the organ of Corti, while loop diuretics such as ethacrynic acid and furosemide and cardiac glycoside such as ouabain affect primarily the vascular stria.

Many drugs have been demonstrated to have receptors in the membranes of the target tissues and frequently these receptors are crucially involved in physiological function, for example transmembrane signaling. There is no reason why the above mentioned ototoxic drugs should be an exception to this rule. This paper describes the latest status of the study in our laboratory to localize the receptors of these ototoxic drugs in the biomembranes of the cochlea.

I. PHOSPHATIDYLINOSITOL BISPHOSPHATE AS RECEPTORS OF POLYAMINO COMPOUNDS

Schacht and coworkers have accumulated evidence to suggest phosphatidylinositol 4,5-bisphosphate ($PhIP_2$) as a receptor for aminoglycoside antibiotics (*30–32*). This lipid is speculated to modulate the plasma membrane permeability by its binding affinity to Ca^{2+} (*6*, *8*). The aminoglycoside antibiotics may displace Ca^{2+} and inhibit calcium-dependent membrane functions. Recently in his laboratory, as well as in ours, it has been shown that other polyamino compounds are likely to share the receptor with aminoglycoside antibiotics (*10*, *41*, *46*).

Polymyxin B, kanamycin A, putrescine and spermine at a concentration of 10 mM in "artificial perilymph" (130 mM NaCl, 10 mM $NaHCO_3$, 4 mM KCl, 1.5 mM $CaCl_2$, 1 mM $MgCl_2$, 5 mM glucose, 20 mM sodium Hepes, 1 mM phosphate buffer, final pH 7.4) were perilymphatically perfused by an infusion pump (Princeton 555) *via* an inlet hole in scala tympani and an outlet hole in scala vestibuli of the basal turn of the cochlea of the guinea pig. The animals were anesthetized with pentobarbital, immobilized with succinylcholine and artificially respirated. The drugs depressed the cochlear microphonics (CM)

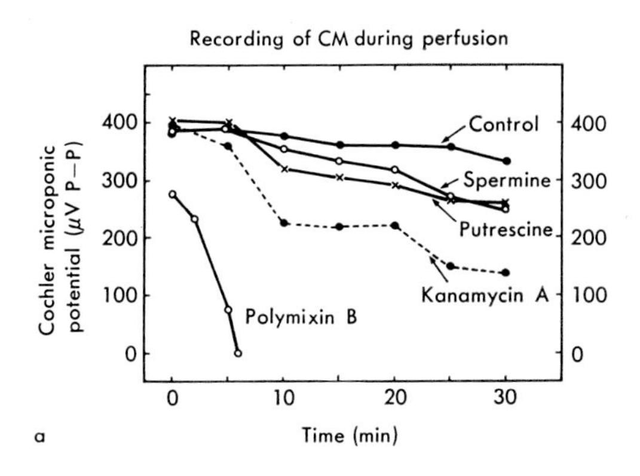

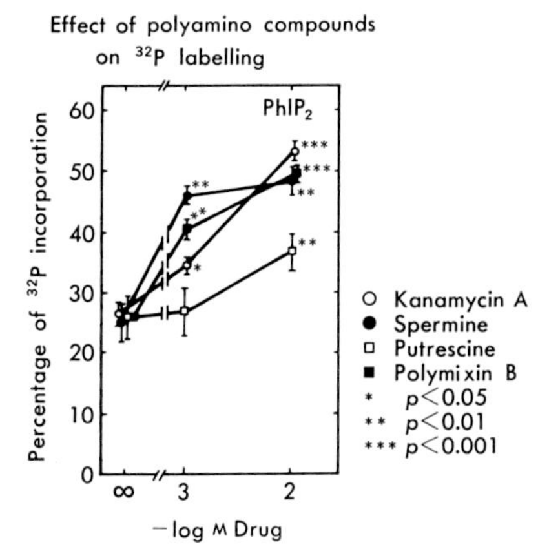

Fig. 1.

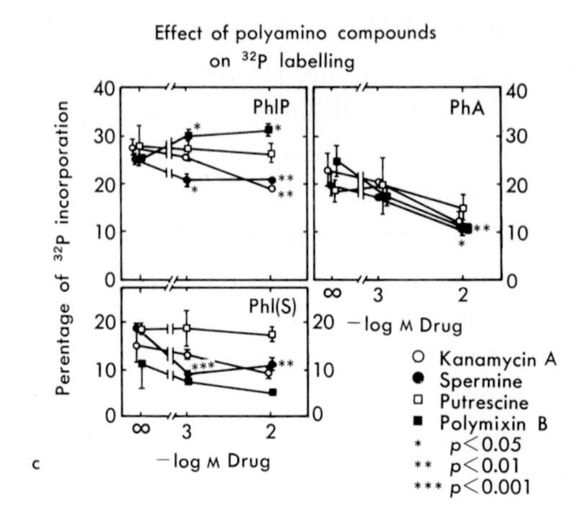

c

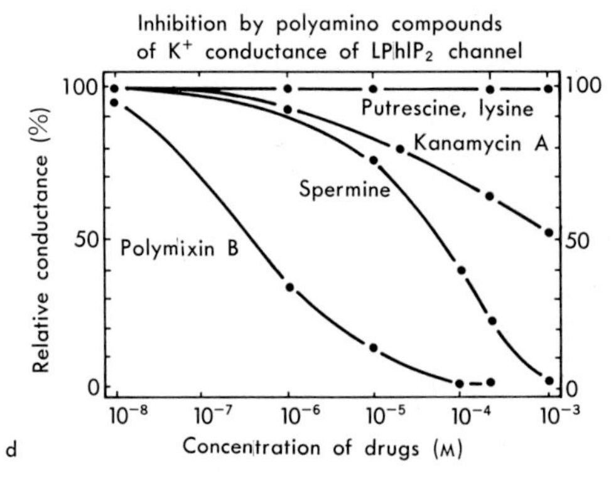

d

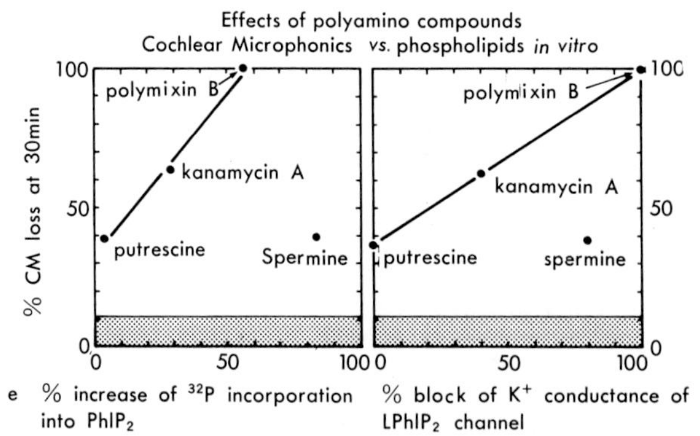

e

Fig. 1. Ototoxicity *in vivo* and effects on polyphosphoinositides *in vitro* of polyamino compounds. a) Effect of polyamino compounds on cochlear microphonic potentials when perilymphatically perfused. Each line represents a perfusion result. b, c) Effect of polyamino compounds on the incorporation of ^{32}P into (b) phosphatidylinositol 4,5-bisphos-

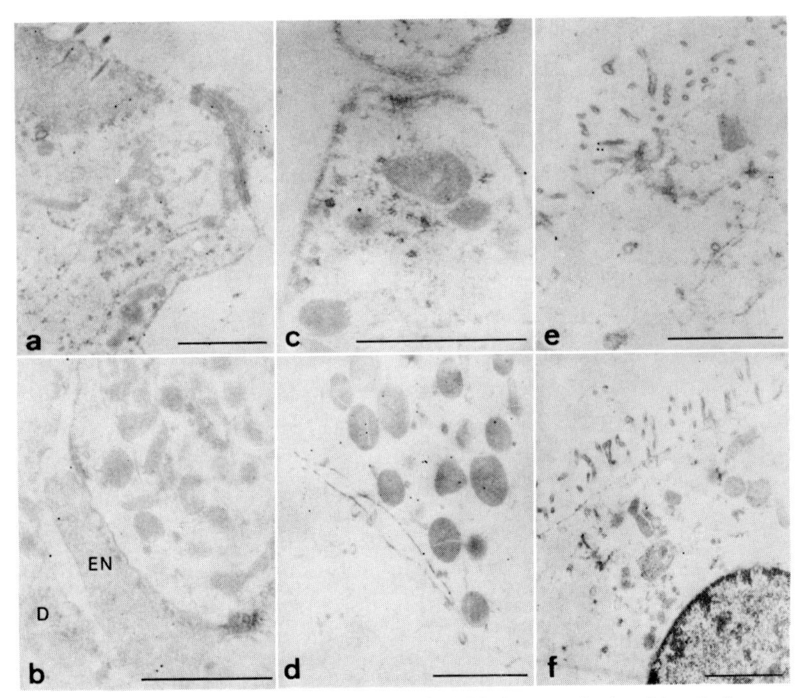

Fig. 2. Localization of the antigenic site of $PhIP_2$ by protein A-gold technique. scale= 1 μm. a–e) Incubation with anti-$PhIP_2$ (diluted 1/400–1/1600). a) Note the concentration of gold particles over the cuticular plate. Particles are also seen on the mitochondrion and plasma membrane of the outer hair cell. b) Note the gold particles over the efferent nerve ending (EN) and Deiters' cell (D). c, d) Gold particles are seen on the plasma membrane and a mitochondrion of the outer hair cell (c), Deiters' cell (d). e) Note gold particles on the microvilli and plasma membrane of Hensen's cell. f) Incubation with anti $PhIP_2$ (diluted 1/800) absorbed with $PhIP_2$. No particles are observed in the (e) region.

recorded from the scala vestibuli of the basal turn in the above described order of toxicity (Fig. 1a).

---

phate ($PhIP_2$), (c) phosphatidylinositol phosphate (PhIP), phosphatidylinositolserine (PhI(s)) and phosphatidic acid (PhA) of brain slice *in vitro*. Values are mean±S.D. of two to five experiments. d) Effect of polyamino compound on the potassium-conductance of lysophosphatidylinositol bisphosphate ($LPhIP_2$) channel. Values were obtained from at least two membranes. e) Relation between ototoxic effects *in vivo* and effects on polyphosphoinositides *in vitro* of polyamino compounds. Ordinate: % loss of cochlear microphonic potential (CM) by 30 min perilymphatic perfusion of 10 mM polyamino compounds. Abscissa: (left) % increase of ^{32}P incorporation into $PhIP_2$ of brain slices when incubated with 10 mM polyamino compounds; (right) % block of potassium conductance of $LPhIP_2$ channel by 200 μM of polyamino compounds.

In another series of experiments guinea pig brain slices (40 mg wt weight) were incubated at 37°C for 60 min in "artificial perilymph" containing 16 μCi of $^{32}P_i$ and drugs (0, 1 or 10 mM). After incubation, lipids were extracted with chloroform-methanol-HCl solution, separated by thin-layer-chromatography and incorporation of $^{32}P_i$ into each lipid fraction was determined by liquid scintillation counting (*22*); $^{32}P_i$ in $PhIP_2$ was most strongly increased (Fig. 1b, c). The magnitude of the effect on $PhIP_2$ correlated well with CM depression by these drugs with the exception of spermine (Fig. 1e, left).

Perilymphatically perfused aminoglycoside antibiotics have been shown to inhibit ^{42}K transport from perilymph to endolymph or to the tissue of vascular stria+spiral ligament (*25*). Sokabe *et al.* recently have shown that lyso-$PhIP_2$ ($LPhIP_2$) in bilayer lipid membranes could form a Ca^{2+}-regulated monovalent cation channel (*34*, *35*); aminoglycoside antibiotics decreased K^+-permeability of this channel (*35*). We examined the effect of "ototoxic polyamino compounds" on the channel. Potassium permeability was depressed in the order of polymyxinB>spermine>kanamycin A>putrescine (Fig. 1d) in good correlation with CM depression except for spermine (Fig. 1e, right). This compound strongly influenced both $^{32}P_i$ incorporation and K^+-permeability, though its depressive effect on CM was limited. Spermine and polymyxin B were demonstrated to be strong inhibitors of gentamicin binding to subcellular fractions of the kidney, while putrescine did not inhibit the binding (*14*).

Although further study is necessary before conclusions can be drawn, the present data suggest that some polyamino compounds share the receptor ($PhIP_2$) with aminoglycoside antibiotics. This suggests that the amino residues of aminoglycoside antibiotics have a crucial role in binding to the receptor.

II. LOCALIZATION OF PHOSPHATIDYLINOSITOL BISPHOSPHATE

Recent progress in immunology and in immunohistochemistry made it possible to raise antibodies to several kinds of lipids and localize them at the ultrastructural level. The antibody to $PhIP_2$ which we used (raised in rabbit; for details refer to (*7*)) reacted virtually only to $PhIP_2$ among the phospholipids tested ($PhIP_2$, phosphatidylinositol phosphate, phosphatidylinositol, phosphatidylcholine, phosphatidylethanolamine, phosphatidic acid) and cholesterol. With it, we localized the antigenic sites of phosphatidylinositol bisphosphate in the cochlea by protein A-gold (pAg) technique (*27*, *28*, *39*).

Young guinea pigs of either sex were used. Immediately after dissection the temporal bone was immersed in 1% glutaraldehyde (from 8% Polyscience) in 0.1 M phosphate buffer, pH 7.4 and the bony capsule was opened to allow further fixation for 2 hr at 4°C. In preliminary experiments, the guinea pig was perfused through the heart with the fixative before immersion of the temporal bone. It appeared that this process did not give sufficient fixation but rather deteriorated morphological preservation and had no beneficial effect on the preservation of antigenicity. Thus the process of arterial perfusion was omitted in later experiments. The cochlea was then rinsed overnight at 4°C with several changes of buffer. After dehydration by a graded series of acetone, the tissue was embedded in Spurr. Ultrathin sections were placed on collodion and carbon-coated nickel grids. After blocking of non-specific binding sites with 1% ovalbumin in 0.1 M phosphate-buffered saline, pH 7.4 (PBS) the sections were reacted with various concentrations of antiserum for 48 hr at 4°C, and then with pAg solution for 30 min at room temperature. Sections were rinsed vigorously with PBS or distilled water between each treatment and were not allowed to be dry throughout the procedure. Finally, the sections were stained with 5% aqueous uranyl acetate solution and observed. The specificity of immunohistochemical labelling was confirmed by negative results obtained when sections were reacted with antisera previously absorbed with $PhIP_2$ or when pAg alone was applied.

In the organ of Corti, antigenic sites to the antibody were detected on the mitochondria and plasma membrane of the hair cells and supporting cells. The cuticular plate, junctional complex, stereocilia, microvilli, synapse and nerve element were also labelled with gold particles (Fig. 2). In the vascular stria, the mitochondria and the plasma membrane of the marginal cell at its basal infolding appeared to be the most heavily labelled sites (Fig. 3a, b).

The localization of $PhIP_2$ in the plasma membrane of the hair cells supports the previously mentioned "$PhIP_2$ hypothesis" for the ototoxicity of aminoglycoside antibiotics. The bound drug may lead to

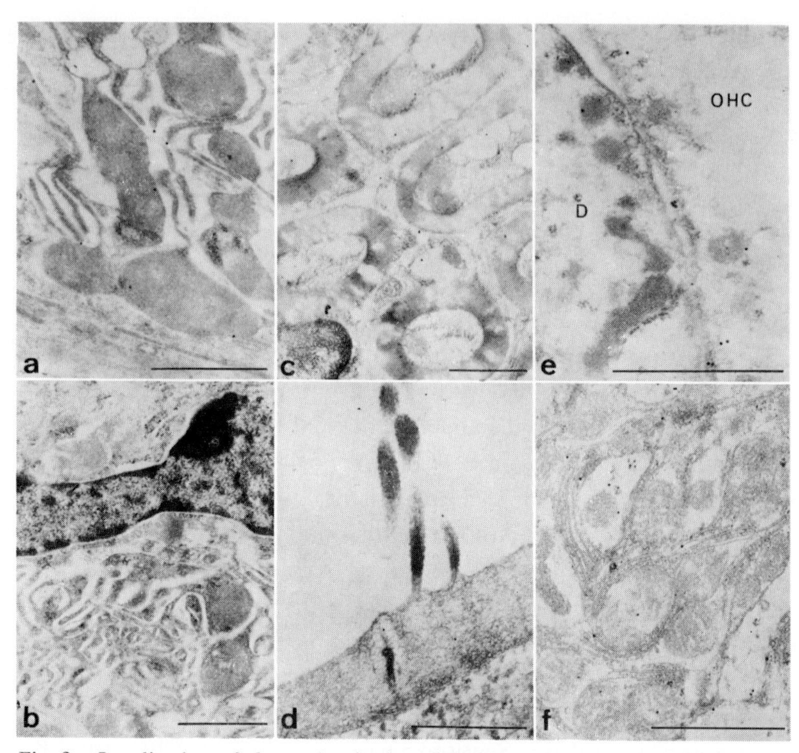

Fig. 3. Localization of the antigenic site of $PhIP_2$ and ethacrynic acid high-sensitive Mg^{2+} ATPase by protein A-gold technique. scale=1 μm. a) Incubation with anti-$PhIP_2$ (diluted 1/1600). Gold particles are observed on the plasma membranes and mitochondria of the basal infolding of the marginal cell. b) Incubation with anti-$PhIP_2$ (diluted 1/800) absorbed with $PhIP_2$. No particles are observed in the (a) region. c–f) Incubation with anti-ethacrynic acid high-sensitive Mg^{2+} ATPase (diluted 1/800–1/2,000). c) Note particles on the myelin of the cochlear nerve. d) Note particles on the stereocilia and cuticular plate of outer hair cell. e) Note particles on the plasma membrane and mitochondria of the outer hair cell (OHC) and Deiters' cell (D). f) Note particles on the plasma membranes and mitochondrion of the basal infolding of the marginal cell.

functional and structural disturbance of the plasma membrane, possibly allowing penetration of the drug into the cell (*45*). The main intracellular binding site of the drug is likely to be mitochondria; some previous morphological and biochemical studies of the cochlea support this notion (*2*, *38*, *39*). In the kidney, gentamicin is demonstrated to bind avidly to mitochondria, microsomal and brush border vesicles *in vitro* (*14*). The greater vulnerability of hair cells than supporting cells to the drug may be explained by the more frequent occurrence of mitochondria in hair cells and the higher metabolic rate. Alternatively, $PhIP_2$ of the plasma membrane of hair cells has a more crucial function than that of supporting cells and the disturbance of its function has a more severe effect. Though the function of $PhIP_2$ in the organ of Corti is still not clearly understood, the uneven localization is suggestive; its high occurrence in the stereocilia and the reticular lamina is of special significance. This area is of great importance for endolymph transduction, and also is the limiting area of peri and endolymph, which consist of quite different electrolytes.

The most active phosphorylation to $PhIP_2$ within the cochlea is known to occur in the vascular stria (*22*). The present study also revealed that the plasma membrane and the mitochondria of the vascular stria, especially at the area of basal infolding is known to contain enzymes such as Na^+K^+-ATPase (*9, 21*), adenylate cyclase (*33, 46*) and ethacrynic acid high-sensitive Mg^{2+}-ATPase (see next section). These enzymes may be important in regulating cochlear fluid and electrolyte balance. $PhIP_2$ in this area may also be involved in this regulation by its binding ability to Ca^{2+}. One may ask then why the vascular stria with abundant aminoglycoside antibiotics receptors escapes toxic involvement. In fact, EP of an animal chronically intoxicated by systemic administration of kanamycin is not affected. The endolymphatic perfusion of kanamycin, however, drastically decreased EP while perilymphatic perfusion did not affect it (*11, 13*). Thus, it is likely that a blood-endolymph or perilymph-endolymph barrier exists to hamper the penetration of aminoglycoside antibiotics into the vascula stria. Ferritin, a cationic substance, cannot pass through the basal membrane of the strial capillary (*29*). Gentamicin is demonstrated to enter endolymph extremely slowly and at a low concentration (*44*).

Perilymphatic perfusion of antibody at an up to two-fold concentration failed to cause any significant change of either CM or EP 150 min after introduction into the cochlea by perfusion (30 min).

III. LOCALIZATION OF OUABAIN-BINDING SITE AND ETHACRYNIC ACID SENSITIVE Mg^{2+}-ATPase

Ouabain, a potent inhibitor of Na^+K^+-ATPase causes drastic reduction of EP when applied perilymphatically (*12, 16, 17*), lowering potassium and raising sodium content in the endolymph (*12*), and reducing the ischemic decline rate of ATP and phosphocreatinine (*17*). A striking similarity between the inhibition of EP by perilymphatic perfusion of ouabain and that of strial Na^+K^+-ATPase by ouabain *in vitro* was also reported (*12*). From this evidence it is suggested that ouabain-sensitive Na^+K^+-ATPase plays a crucial role in the vascular stria, possibly secreting potassium into and absorbing sodium from the endolymph.

In some tissue the Na^+/K^+ pump (Na^+K^+-ATPase) has been successfully localized by autoradiographic technique using ^{3}H-ouabain (*3, 36*); therefore this technique was applied to the cochlea in order to localize the ouabain binding site (*21*). Ouabain at a concentration of 10^{-5} M in artificial perilymph was perfused perilymphatically, as described earlier, at the rate of 10 μl/min while monitoring EP by a glass electrode inserted into the scala media through the vascular stria. The cochlea was then dissected for electron microscopic or autoradiographic observation. In the latter case ^{3}H-ouabain (180 Ci/ml, final specific activity 18 Ci/mmol) was included in the perfusate. The tissue was frozen in isopentane precooled to −90°C, freeze-dried *in vacuo* at −40°C for 96 hr, fixed with osmium gas and acrolein vapor and embedded in Spurr-resin *in vacuo*. The 1 μm sections were cut, covered with emulsion (Sakura NR-M2) and exposed for 4 weeks; after development, they were stained with toluidine blue.

The EP was depressed by ouabain, became negative, reached a minimal value within 40 min and then slowly recovered but still remained negative at 60 min of perfusion (Fig. 4a). The autoradiographic localization of ouabain was performed at several time points and localization was essentially the same throughout the 60 min perfusion. It was surprising that even after 60 min of perfusion there was a higher concentration of ouabain in the spiral ligament around the region of spiral prominence than in the vascular stria (Fig. 4b). Within the vascular stria the lowermost area adjacent to spiral prominence was the one most densely labelled (Fig. 4c). Closer examination of this area revealed that it was the marginal cells, especially their basal infoldings, where ouabain sensitive Na^+K^+-ATPase was clearly located by histochemistry (*9, 20*). In the organ of Corti, the havenula perforata and pillar cells were the most heavily labelled areas (Fig. 4d), suggesting a penetration

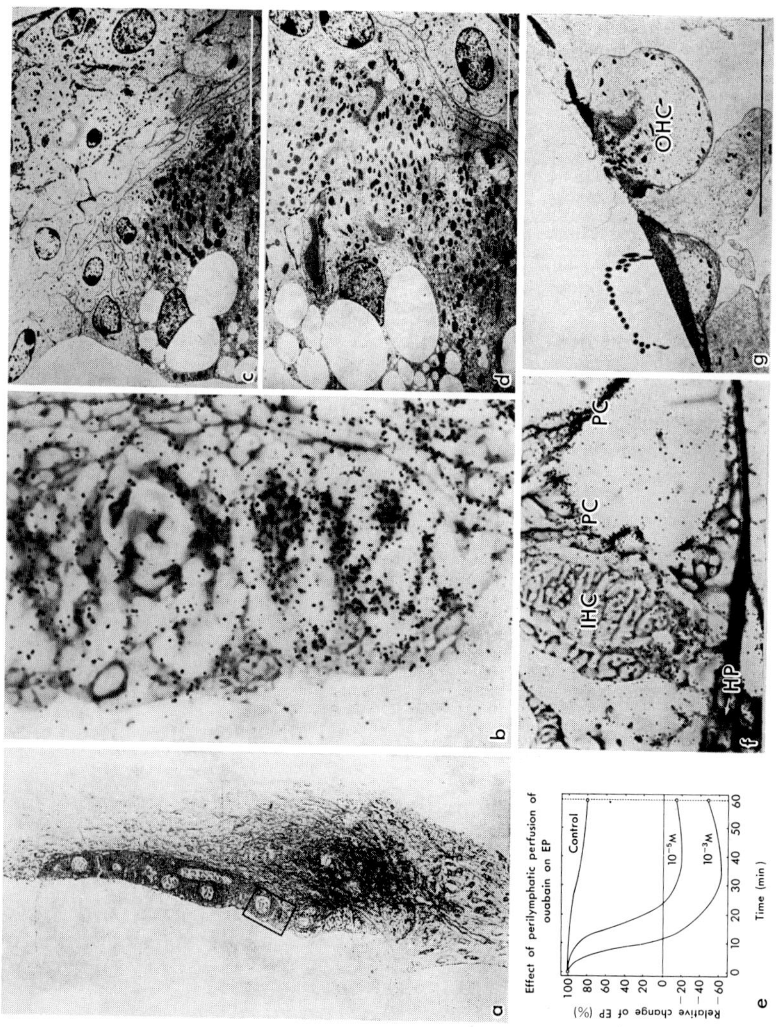

Fig. 4. Distribution and effects of perilymphatically perfused ouabain. Duration of the perfusion was 60 min. a) Autoradiograph (ARG) using ^{3}H-ouabain showing the lateral wall of the cochlear duct. Note the silver grains accumulated in the spiral ligament adjacent to spiral prominence. b) ARG of ^{3}H-ouabain: enlargement of the lowest area of the vascular stria indicated by the rectangle in (a). The accumulation of grains on the basolateral membrane of the marginal cell was observed only in this area within the vascular stria. c) Electron micrograph (EM). Lowest area of the vascular stria (VS) after 60 min of 10^{-3} M ouabain perfusion. Note the intense vacuolization of the marginal cell. scale=10 μm. d) EM. Uppermost area of VS after 60 min of 10^{-3} M ouabain perfusion. Vacuolization of marginal cells is as intense as in the lowest area shown in (c). scale=10 μm. e) Endocochlear potential (EP) during perfusion. EP is depressed by

route of the tracer from perilymph to cortilymph. No specific accumulation of silver grains was observed at the hair cells themselves.

Morphological examination after ouabain treatment showed intense and moderate vacuolization in all marginal and intermediate cells regardless of their distance from the spiral prominence. Histopathological changes in the spiral ligament and the epithelium of the spiral prominence were minimal. In contrast, the outer hair cells were affected and ballooning of cytoplasm through the cuticular plate was observed.

Thus, an apparent discrepancy exists between the localization of ouabain binding and affected cells. These data, however, may be consistent if some speculations are allowed. Perilymphatically perfused ouabain cannot penetrate directly into the vascular stria because of the perilymph-endolymph barrier (there are extremely well developed tight junctions between basal cells (*26*, *37*)) and accumulates first where Na^+K^+-ATPase activity is highest within the perilymphatic space (around the spiral prominence) (*15*). If such a barrier is bypassed, *e.g.*, by incubation *in vitro*, more ouabain accumulates in the vascular stria than in the spiral ligament (*9*). However, once a marginal cell is affected and its intracellular ionic balance disturbed, this disturbance may spread throughout the vascular stria, since all of its component cells are ionically connected by gap junctions (*26*, *37*). The relative resistance of the spiral ligament cell to ouabain may be due to its less active energy metabolism (it blocks energy utilization). The histopathological change of hair cells seen at the cuticular lamina may be due to the change in ionic concentration of the endolymph (increase of sodium and decrease of potassium) (*13*).

Ethacrynic acid, a potent loop diuretic, causes drastic reduction of EP when systemically or perilymphatically applied (*17*, *23*). Among several proposed receptor candidates of this drug in the cochlea were Na^+K^+-ATPase and adenylcyclase (*17*), but recent evidence suggests that neither enzyme is a primary site of pharmacological action in this organ (*17*, *19*, *43*). Ethacrynic acid high-sensitive Mg^{2+}-ATPase ac-

---

ouabain dose-dependently. f) ARG using ^3H-ouabain: organ of Corti. Grains were observed on the habenulla perforata (HP) and pillar cells (PC), but such an accumulation was not observed on the inner hair cells (IHC). g) EM. Organ of Corti. Note protrusion of the cytoplasma of an outer hair cell (OHC) through the cuticular plate. scale=10 μm.

tivity was recently found in microsomal fractions from cortical gray matter of the rabbit brain (*4, 5*). This activity is reduced by replacement of Cl^- with several anions which are likely to be inhibited by ethacrynic acid reacting with thiols and associated with plasma membranes and/or endoplasmic reticula (*4, 5*). Using the antibody to the enzyme (raised in Wistar rats), antigenic sites of ethacrynic acid high-sensitive Mg^{2+}-ATPase were localized in the cochlea by pAg technique (*27, 28*). Preliminary results are shown here.

Young Wistar rats of either sex and the method discussed in the previous section were used. The most consistent labelling with gold particles was observed on the myelin of the cochlear nerve (Fig. 3c). In the organ of Corti, the stereocilia, plasma membranes and mitochondria of the hair and supporting cells were labelled with gold particles (Fig. 3d, e). In the vascular stria, antigenic sites were detected frequently in the plasma membranes and mitochondria of the marginal cell at its basal infolding (Fig. 3f).

It is of interest that ethacrynic acid high-sensitive Mg^{2+}-ATPase was located in the region of the vascular stria where adenylcyclase and Na^+K^+-ATPase (*9, 20*) were observed. The significance of this finding and the occurrence of the enzyme in the organ of Corti require further study.

The occurrence of the antigenic site in mitochondria was unexpected since the enzyme has been determined to be of non-mitochondrial origin (*4*). Cross antigenicity of the used antibody to ATPase of mitochondrial origin is possible.

SUMMARY

The receptors of some ototoxic drugs have been identified. Phosphatidylinositol bisphosphate is the most likely receptor for aminoglycoside antibiotics; in the present study it was shown that other ototoxic polyamino compounds may share this receptor. A kind of Mg^{2+}-ATPase is possibly one of the receptors for ethacrynic acid. These receptors were localized in the cochlea using an immunohistochemical method, the pAg technique. Ouabain binding sites were localized by autoradiographic labelling.

In the vascular stria, the basolateral plasma membrane of the marginal cell (basal infolding) was enriched by the proposed receptors for all three kinds of ototoxic drugs. Functionally these receptors might be involved in the control of the balance of electrolytes (Ca^{2+}, Na^{+}, K^{+} and Mg^{2+}) and cochlear fluids. The lack of aminoglycoside antibiotic toxicity to this region in spite of the existence of the receptor may be explained by inaccessibility of the drug to the region.

In the organ of Corti, these receptors are distributed unevenly suggesting distinct roles in auditory transduction. However, the selective susceptibility of hair cells over supporting cells to ototoxic drugs did not seem to be explained by the distribution of the receptors alone. In this respect, intracellular receptors, perhaps in mitochondria, might also be involved.

Acknowledgments

The author gratefully acknowledges the collaboration of Dr. T. Yoshioka and Prof. C. Inagaki (both of whom also donated antibodies for the experiments), and Drs. M. Sokabe, M. Machino, H. Morioka, W. Oshima, S. Komiya, F. Mizukoshi and C. Yamada. The author also gratefully acknowledges the critical reading of this manuscript by Dr. J. Schacht. Thanks are also due to Miss R. Sarvis for her assistance with the manuscript.

Some parts of this article will be published as original papers.

REFERENCES

1 Davies, H. (1965). *Cold Spring Harbor Symp. Quant. Biol.* **30**, 181–190.
2 Duvall, J.A. and Wersall, J. (1963). *Acta Otolaryngol.* **65**, 581–598.
3 Ernst, S.A. and Mills, J.W. (1980). *J. Histochem. Cytochem.* **28**, 72–77.
4 Hara, M., Fujiwara, M., and Inagaki, C. (1982). *Biochem. Pharmacol.* **31**, 4077–4079.
5 Hara, M., Miwa, S., Fujiwara, M., and Inagaki, C. (1982). *Biochem. Pharmacol.* **31**, 877–879.
6 Hendrickson, H.S. and Reinertsen, J.L. (1971). *Biochem. Biophys. Res. Commun.* **44**, 1258–1264.
7 Horikoshi, T., Yanagisawa, K., and Yoshioka, T. (1984). *Proc. Japan. Acad.* **60B**, 157–160.
8 Kai, M. and Hawthorn, J.H. (1969). *Ann. N.Y. Acad. Sci.* **165**, 761–773.
9 Kerr, T.P., Ross, M., and Ernst, S.A. (1982). *Am. J. Otolaryngol.* **3**, 332–338.
10 Komiya, S., Ono, Y., Mizuta, W., Oshima, W., and Tachibana, M. (1982). *Ear Res. Jpn.* **13**, 100–101 (in Japanese).
11 Komune, S. and Snow, J.B. (1981). *Otolaryngol. Head Neck Surg.* **89**, 1013–1018.
12 Konishi, T. and Mendelsohn, M. (1970). *Acta Otolaryngol.* **69**, 192–199.
13 Konishi, T. (1979). *Acta Otolaryngol.* **88**, 41–46.

14 Kornguth, M.L., Bayer, W.H., and Kunin, C.M. (1980). *J. Antimicrob. Chemother.* **6**, 121–131.
15 Kuijpers, W. and Bonting, S.L. (1969). *Biochem. Biophys. Acta* **173**, 477–485.
16 Kuijpers, W. and Bonting, S.L. (1970). *Pflugers Arch.* **320**, 348–358.
17 Kusakari, J., Ise, I., Comegys, T.H., and Thalmann, R. (1978). *Laryngoscope* **88**, 12–37.
18 Lodhi, S., Weiner, N.D., Mechigian, I., and Schacht, J. (1980). *Biochem. Pharmacol.* **29**, 597–601.
19 Marks, S.C. and Schacht, J. (1981). *Scand. Audiol.* **S 14**, 131–138.
20 Mess, K. (1983). *Acta Otolaryngol.* **95**, 277–289.
21 Mizukoshi, F., Tachibana, M., Oshima, W., Nishimura, H., Yasuda, N., Machino, M., Morioka, H., and Mizukoshi, O. (1983). *Ear Res. Jpn.* **14**, 126–132 (in Japanese).
22 Orsulakova, A., Stockhorst, E., and Schacht, J. (1976). *J. Neurochem.* **26**, 285–290.
23 Prazma, J., Thomas, W.G., Fischer, N.D., and Preslar, M.J. (1972). *Arch. Otolaryng.* **95**, 448–456.
24 Quick, C.A. and Duvall, A.J. (1970). *Laryngoscope* **80**, 954–965.
25 Rauch, S. and Rauch, I. (1974). In *Handbook of Sensory Physiology*, vol. V/1, ed. Keidel, W.D. and Neff, W.D., pp. 647–682. Berlin, Heidelberg, New York: Springer-Verlag.
26 Reale, E., Luciano, L., Franke, K., Pannese, E., Wermbter, G., and Iurato, S. (1975). *J. Ultrastructure Res.* **53**, 284–297.
27 Roth, J., Bendayan, M., and Orci, L. (1978). *J. Histochem. Cytochem.* **26**, 1074–1081.
28 Roth, J. (1982). In *Techniques in Immunocytochemistry*, vol. 1, ed. Bullock, G.R. and Petrusx, P., pp. 107–133, London, New York, Tokyo: Academic Press.
29 Santos-Sacchi, J. (1980). *Acta Otolaryngol.* **89**, 12–26.
30 Schacht, J. (1974). *Ann. Otol. Rhinol. Laryngol.* **83**, 613–618.
31 Schacht, J. (1976) *J. Neurochem.* **27**, 1119–1124.
32 Schacht, J. (1979). *Arch. Otorhinolaryngol.* **224**, 129–134.
33 Schacht, J. (1982). *Am. J. Otolaryngol.* **3**, 328–331.
34 Sokabe, M. (1982). *Bull. Fac. Human Sci. Osaka Univ.* **8**, 241–288.
35 Sokabe, M., Hayase, J., and Miyamoto, K. (1982). *Proc. Japan Acad.* **58** (Ser. B), 177–180.
36 Stirling, C.E. (1972): *J. Cell Biol.* **53**, 704–714.
37 Tachibana, M. (1977). *Jpn. J. Otol.* **80**, 39–50 (in Japanese).
38 Tachibana, M. (1977). *Jpn. J. Otol.* **80**, 55–80 (in Japanese).
39 Tachibana, M. (1980). *Histochemistry* **81**, 157–160.
40 Tachibana, M., Anniko, M., and Schacht, J. (1982). *Acta Otolaryngol.* **96**, 31–38.
41 Takada, A., Lodhi, S., Weiner, N.D., and Schacht, J. (1982). *J. Pharma. Sci.* **71**, 1410–1411.
42 Tamura, H. (1978). *Audiology Jpn.* **21**, 668–687 (in Japanese).
43 Thalmann, I., Paloheimo, S., and Thalmann, R. (1981). *J. Acoust. Soc. Am.* **69**, S 112.
44 Tran Ba Huy, P., Manuel, C., Meulemans, A., and Sterkers, O., and Amiel, C. (1981). *J. Infect. Dis.* **143**, 476–486.
45 Weiner, N.D. and Schacht, J. In *Aminoglycoside Ototoxicity*, ed. Lerner, S.A., Matz, G.J., and Hawkins, J., Jr., pp. 113–121. Boston: Little, Brown.
46 Yamada, C., Komiya, S., and Tachibana, M. (1983). *Ear Res. Jpn.* 227–229 (in Japanese).
47 Zajic, G., Anniko, M., and Schacht, J. (1983). *Hearing Res.* **10**, 249–262.

10

AMINOGLYCOSIDE BLOCKADE OF CATION CHANNELS FROM POLYPHOSPHOINOSITIDES AND SARCOPLASMIC RETICULUM IN PLANAR BILAYER MEMBRANES

MASAHIRO SOKABE

Department of Behaviorology, Faculty of Human Sciences, Osaka University, Osaka 565, Japan

Ion channels have crucial roles in transmembrane signalling, however the molecular mechanism of the control of ion permeation through channels remains one of the greatest missing links in the stimulus-response coupling of excitable cells. A certain class of phospholipids which includes phosphoinositides and phosphatidic acid itself has been suggested to serve as an ionophore or a regulator of ion permeation (*9, 22, 26, 34*). It has repeatedly been speculated that polyphosphoinositides (PPI), phosphatidylinositol-4,5-bisphosphate($PhIP_2$) and phosphatidylinositol 4-phosphate(PhIP), have a key role in the control of Na^+, K^+ permeability of excitable membranes (*1, 15, 19, 37*) due to their unique bio- and physicochemical properties: 1) Although PPI are trace lipids in eukaryotic cells, they may be concentrated at plasma membranes (*23*). 2) They show rapid phosphorus turnover which may be accelerated by appropriate electrical or chemical stimulation to the excitable tissues (*1, 4, 5, 8, 38, 39*). 3) PPI are water soluble lipids having an extremely high affinity for calcium ion, and they show hydrophilic to hydrophobic transition with binding of Ca^{2+} (*14, 33*). Based on these properties, Kai and Hawthorne (*19*) originally proposed that PPI may control monovalent cation permeability of excitable membranes through a regulation of membrane bound Ca^{2+}. More recently, following their hypothesis, we examined the effect of PPI on the ionic permeability of planar bilayer membranes and found that lysoPPI themselves formed Ca^{2+} sensitive monovalent cation permeable channels in the bilayer although intact PPI had no effect (*12*, *33*).

Aminoglycoside (AG(s))antibiotics have been widely used as therapeutic drugs but, unfortunately, cause side effects such as nephrotoxicity or ototoxicity (*28*); their underlying mechanism is still unknown. PPI have been suggested to be a primary receptor for the drug because the latter specifically inhibits PPI metabolic turnover in the inner ear probably by a direct binding to PPI (*29*). In acute ototoxicity, AGs inhibited the receptor potentials from hair cells (*17*, *18*). Since these potentials originated from the permeability increase of the monovalent cation channel on hair cell membranes (*6*), it is possible that the reduction of receptor potentials by AGs is caused by a blockade of the cation channel. Using these findings and suggestions, we investigated the effects of AGs on the lysoPPI channel to test the latter's applicability for modeling mechanisms of AGs acute ototoxicity, and succeeded in simulating several aspects of the ototoxicity (*35*).

Because of the difficulty of single channel recordings from the sound-transduction channel of hair cells, whether or not AGs directly affect the channel is unknown; in fact, no direct interaction between AGs and biological channels has yet been established. To clarify this, the mode of action of AGs on the ion channels should be analyzed at the single channel level. Using a K^+ channel from fragmented sarcoplasmic reticulum (SR) incorporated in planar bilayers, we investigated the mode of AGs action on the channel and showed that AGs behave as potent channel blockers (*34*). Interestingly, the mode of action of AGs on the SR-K^+ channel resembles that on the lysoPPI channel.

In this article, I will describe the basic properties of the lysoPPI channel, AGs effects on it as a model for acute ototoxicity, and finally the channel blockade of SR-K^+ channel by AGs, and will discuss the similarities of the two different kinds of ion channels in planar bilayers.

I. MONOVALENT CATION CHANNEL FROM POLYPHOSPHO-INOSITIDES (PPI)

The affinity for Ca^{2+} of PPI is significantly decreased with the hydrolysis of monoester phosphates in the order, $PhIP_2 > PhIP > PhI$ (*15*). It has been suggested that interconversion of these lipids may control the Na^+, K^+ permeability of excitable membranes *via* binding and release of Ca^{2+} (*15, 19*). However, this hypothesis seems to have hardly been tested directly on biological membranes because of their molecular complexity. Most of the past studies have provided only correlations of the metabolic turnover of these lipids with the membrane permeability changes, and have given no direct evidence indicating an alteration of membrane permeability by the lipids. So, I would propose here a simpler working hypothesis: "$PhIP_2$ may form an ionic channel with a conductance which is regulated by Ca^{2+}", which can be tested directly on planar bilayer membranes. Following this line, we tried to incorporate $PhIP_2$ into planar bilayers. The intact $PhIP_2$, however, had no effect on the membrane permeability, but a trace amount of lyso$PhIP_2$ drastically decreased the membrane resistance by forming stable monovalent cation permeable channels (*12, 33*).

Lyso$PhIP_2$ was prepared by autoxidation (*12*) or enzymatically cleaved by phospholipase A_2 of $PhIP_2$ from bovine brain; the resulting samples gave a single spot on a silica gel 60HPTLC plate (Merck). Sometimes the crude samples were further purified on a DEAE cellulose column (Whatman DE 32) to remove free fatty acids and lipid peroxides. The purified and crude samples gave a single spot at the same position on the HPTLC plate, and the molar ratio of the ester fatty acid to phosphorus of the samples was determined to be 1:3, while that of intact $PhIP_2$ was 2:3. All crude and purified samples showed basically the same effect on the monovalent cation permeability of bilayers suggesting that an active component of our samples was lyso$PhIP_2$. Planar bilayers were formed from a neutral lipid such as oxidized cholesterol or glycerylmonooleate to get rid of the surface charge effect. After thinning of the membrane, concentrated lyso$PhIP_2$ and Ca^{2+} were added to both sides of the membrane under constant stirring. Within 5 min, the lysoPhIP$_2$-Ca^{2+} complexes were incorporated into the bilayer to form ion channels.

The basic properties of this ion channel are as follows: 1) It is selectively permeable to monovalent cations. Anions and polyvalent cations were scarcely permeable compared to monovalent cations. Selectivity sequence judged from the permeability coefficient ratio was NH_4^+ (1.3)$>$$K^+$ (1)$>$$Rb^+$ (0.9)$>$$Cs^+$ (0.8)$>$$Na^+$ (0.6)$>$$Li^+$ (0.3), tetramethylammonium (0.3)$\gg$tetraethylammonium (0.01) (*11*). From this sequence, the diameter of the channel could be estimated as at least less than 7 Å. 2) Divalent cations like Ca^{2+} are not only impermeable ions but also reversible inhibitors to the channel. This inhibition was mainly due to a specific interaction with the channel such as channel blocking or gating, but not to a nonspecific screening effect (*33*). 3) The ion permeable system was only formed when lysoPhIP$_2$ was added to both sides of the membrane. Addition of the lipid to only one side induced no effect, suggesting that lysoPhIP$_2$(s) forms a so-called half pore (*33*) on a half leaflet of a bilayer and that an association of the half pores on each side of the bilayer formed a completed cation channel spanning the bilayer. This also suggests that lysoPhIP$_2$ does not translocate itself from one side to the other of the membrane as does a mobile ion carrier. In addition, the ion permeable system of lysoPhIP$_2$ was irreversibly incorporated in the planar bilayers. LysoPhIP could also form a cation channel similar to lysoPhIP$_2$ except that its affinity for Ca^{2+} or AGs is much lower. These results strongly suggest that lysoPPI can form Ca^{2+}-sensitive monovalent cation channels in planar bilayers. Recently we were able to measure the putative single channel currents probably originating from the lysoPhIP$_2$ channel on both oxidized cholesterol and glycerylmonooleate membranes. The single channel conductances were identical, 6pS in the presence of 500 mM KCl, a magnitude which is excessive for a carrier mechanism (*16*, *21*).

II. AMINOGLYCOSIDE EFFECTS ON LysoPPI CHANNEL AND OTOTOXICITY

Although lysoPPI formed cation channels in planar bilayers it is unclear whether this lipid acts as a channel in biological membranes. We took a biochemical approach to this problem (*13*) but a quantitative analysis of lysoPPI seemed rather difficult mainly due to their extremely limited quantity in biological membranes. Thereupon, we decided to evaluate the lysoPPI channel from another point of view: to what degree does the channel resemble biological channels. As an initial step, some pharmacological properties of this channel were studied. A K^+ channel blocker (tetraethylammonium), Na^+ channel blocker (tetrodotoxin), nicotinic agonist (carbamylcholine), nicotinic antagonist (α-bungarotoxin), and muscarinic antagonist (atropine), all had slight effect on the K^+-permeability of the lysoPPI channel. As mentioned

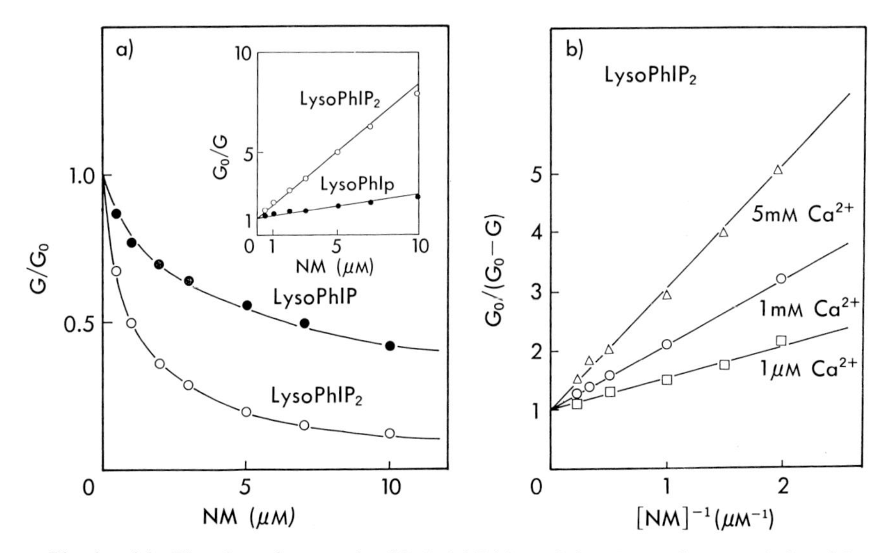

Fig. 1. (a) Titration of neomycin (NM) inhibition of the K^+-conductance induced by lysoPhIP$_2$ and lysoPHIP on oxidized cholesterol membranes. The ordinate indicates relative conductance defined as a ratio of the conductance at a given concentration of NM (G) and that in the absence of NM (G_0). Aqueous phase was composed of symmetrical solutions of 100 mM KCl, 1 mM $CaCl_2$ and 1 μM lysoPhIP$_2$ or lysoPhIP. Concentrated NM solution was successively added to one side of the membrane. Membrane conductances were obtained by measuring the currents with the application of +10 mV. Data were taken from at least two different membranes. Inset: reciprocal plot of equation $G/G_0 = \{1+[NM]/K_d\}^{-1}$ which indicates a single site titration curve, where K_d means dissociation constant. Each point nearly follows the theoretical lines ($K_d = 1.25$ μM and 7 μM for lysoPhIP$_2$ and lysoPhIP, respectively) suggesting that one NM molecule blocks one ion channel. (b) Double reciprocal plot of the inhibition rate ($[G_0 - G]/G_0$) *vs.* NM concentration in the presence of various concentrations of Ca^{2+}; 1 μM Ca^{2+} (□); 1 mM Ca^{2+} (○); 5 mM Ca^{2+} (△). The data gave straight lines crossing a common intersection on the ordinate indicating that Ca^{2+} was a competitive inhibitor to the NM blockade of PhIP$_2$ channel. The aqueous phase is the same as in Fig. 1a except for Ca^{2+} concentration.

above, however, divalent cations like Ca^{2+} and AGs markedly inhibited the K^+-permeability of the channel.

AGs are well-known for their toxic actions to the kidney, inner ear, vestibular organ and neuromuscular junctions (*28*), as well as for their antibacterial action. Although the mechanisms of actions of the drugs are still unclear, certain toxicities to excitable tissues have been suggested to occur at the membrane surface (*29*), because strong basic AGs do not readily permeate the cell membranes (*27*) and an accumulation of the drug has been shown in the membranes (*3*). The relation between AGs toxicity and ion channels may be inferred from the studies on the acute ototoxicity. As mentioned in the introduction, AGs acutely reduced the hair cell receptor potentials originating from permeability changes of the monovalent cation channels on hair cell membranes (*6*). The simplest interpretation of the receptor potential reduction by the drug is that the drug interacts directly with the channel to block the cation permeation. On the other hand, Schacht and his collegues (*31*) have provided several pieces of evidence suggesting that PPI may be a primary receptor-molecule for AGs. In addition, lysoPPI formed a cation channel in bilayer membranes. With these findings and suggestions we investigated the effects of AGs on the lysoPPI channel to test its applicability as a model for the acute ototoxicity in hair cell membranes.

As shown in Fig. 1a, neomycin (NM) reduced the K^+-conductance of the lysoPPI channel with an apparent dissociation constant of approximately 1.25 μM and 7 μM for the lysoPhIP$_2$ and lysoPhIP channels, respectively. This inhibition followed the single site titration curve:

$$G/G_0=\left\{1+\frac{[\mathrm{NM}]}{K_d}\right\}^{-1} \tag{1}$$

where G/G_0 is the relative conductance and K_d is the apparent dissociation constant. If we assume that NM blocks the channel, it is suggested that one NM molecule blocks one lysoPPI channel. Next, the effect of Ca^{2+} on the NM effect was investigated, because Ca^{2+} is known to reduce the acute ototoxicity of AGs (*10*); it was found that Ca^{2+} decreased the NM effect in a dose dependent manner. Figure 1b shows a double reciprocal plot of inhibition rate ($(G_0-G)/G_0$) *vs.* NM concentration in the presence of various concentrations of Ca^{2+}. From this figure, it can be concluded that Ca^{2+} competitively inhibited the NM effect on the lysoPPI channel. Conversely, a small amount of NM competitively displaced the lysoPhIP$_2$ channel-bound Ca^{2+} (*35*). Finally, the effect of several AGs was compared. The drugs also reduced the K^+-conductances in the same manner following the order: tobramycin, gentamicin, neomycin$>$kanamycin, fortimycin$>$streptomycin, ribostamycin. The order closely parallels the ototoxic actions of these AGs (*2*). Moreover, the selectivity pattern among univalent cations (see previous section) resembles that of the hair cell transduction channel (*6*).

To conclude, the lysoPhIP$_2$ channel apparently can simulate several aspects of the AGs acute ototoxicity and, at the same time, it may offer a molecular- and bioelectric-level model for understanding an important part of the sound transduction by hair cells. Thus, if this channel operates as a sound transduction channel, several aspects of AGs acute ototoxicity may be explained. If so, AGs exert their toxicity by a direct blocking of the channel and/or displacement of Ca^{2+} from the lysoPPI channel. Schacht recently proposed that the metabolic turnover of PPI in hair cells may not correlate with sound transduction (*30*). The sound transduction may instead be carried out by changes in the physical properties of the lysoPPI channel through unknown mechanisms, probably including the Ca^{2+} regulating process which is indispensable for the generation of receptor potentials (*6*). As has been mentioned, lysoPhIP also formed a cation channel similar to lysoPhIP$_2$, but since the affinity for AGs of lysoPhIP is lower than that of lysoPhIP$_2$ (Fig. 1a), the primary receptor for AGs may be (lyso)PhIP$_2$.

III. AMINOGLYCOSIDE EFFECT ON A K^+ CHANNEL FROM SARCOPLASMIC RETICULUM

In the previous section, I discussed the mechanism of AGs acute ototoxicity by assuming that AGs may block the monovalent cation channel which serves as a sound transduction channel at the hair cells. However, it is still unclear whether AGs block this channel on hair cell membranes. Since the possibility of an AGs blockade of the channel has not been seriously considered, there is actually no direct evidence that AGs block any biological channels. A single channel experiment should be done to clarify this, but single channel recordings from hair cells have not yet been performed because of technical difficulties. Therefore, we decided to investigate, at the single channel, whether AGs act as a blocker to certain cation channels of biological membranes. Recently, two kinds of single channel recording techniques, the patch clamp and planar bilayer methods, have been developed (*20*). We employed a K^+ channel from fragmented SR incorporated in a planar bilayer because the single channel conductances of SR-K^+ channel (*e.g.*, 160 pS at 100 mM K^+) are relatively large and because the planar bilayer system makes it easier to analyze the detailed mode of drug actions to the channel and to control the ionic composition of aqueous phases.

SR vesicles from rabbit skeletal white muscle were incorporated into planar bilayers of asolectin by the Ca^{2+}-induced fusion process which, as described by Miller (*24*), is mainly dependent on the osmotic gradient across SR vesicle membranes. After thinning of the planar membrane, Ca^{2+} (0.5–1 mM, final concentration) and SR vesicles (5–20 μg protein/ml, final concentration) were added to one side of the membrane (the *cis* side) with constant stirring. Within a few minutes after the SR addition, the membrane conductance began to increase in discrete steps which reflected fusion of SR vesicles with the bilayer. For measurement of macroscopic conductances, fusion was stopped simply by allowing enough time to pass so that the SR vesicles dissipated their osmotic gradient. When a low concentration of SR (1–5 μg/ml) was added, sometimes only a single SR vesicle containing a few channels would fuse and opening-closing fluctuations of individual channels could be observed. Since the SR vesicle contains both K^+ and Cl^- channels, K^+-gluconate buffer was used for the K^+-conductance measurement, because K^+ and gluconate$^-$ do not permeate the Cl^- channel. The side to which SR was not added (the *trans* side) was defined as virtual ground.

First, NM effects on the macroscopic K^+ conductance were investigated. Upon addition of NM to the *cis* side, the K^+ conductance reversibly reduced, whereas Cl-conductances were unchanged (*34*). *Trans* addition of NM also reduced the K^+ conductance in a similar way. One of the most feasible mechanisms of this conductance inhibition may be surface potential changes by the drug. Since the bilayer carries a negative surface charge, a cationic drug, NM will reduce the magni-

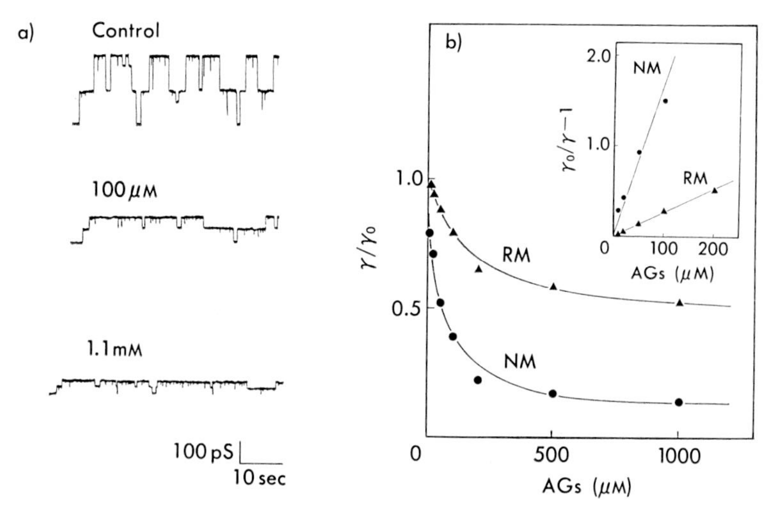

Fig. 2. (a) Single channel recordings in the presence of various concentrations of NM in the *cis* side at +50 mV. The single channel conductance was gradedly reduced with the increase of NM accompanied by an elongation of the channel open time, suggesting that NM behaves as an open channel blocker to the SR-K^+ channel. The aqueous phase was 5 mM HEPES-Tris (pH 7.2), 1.2 mM EDTA, 1 mM $CaCl_2$, 100 mM K^+ gluconate, 5 μg SR/ml. All recordings were made in different membranes. (b) Aminoglycosides (●, neomycin; ▲, ribostamycin) titration of single channel conductance of SR-K^+ channel in the presence of drugs in the *cis* side only. Aqueous phase was the same as in (a). Inset: plot of $[(\gamma_0/\gamma)-1]$ against AGs concentration where γ is the channel conductance at a given concentration of AGs, γ_0 is the channel conductance in the absence of the drug. The solid lines were drawn according to $[(\gamma_0/\gamma)-1]=[\mathrm{AGs}]/K_b(V)$ with K_b (50 mV) = 60 μM and 400 μM for neomycin and ribostamycin, respectively. Data were taken from at least three membranes.

tude of the surface potential by a charge screening or a specific binding, which will reduce the local concentration of K^+ near the mouth of the channel. However, this possibility seems ruled out, because Cl-conductances measured under the same conditions were not affected by NM. Therefore, another mechanism, a channel blockade, or an inhibition of the channel opening reaction must be considered. These possibilities were tested by single channel experiments. Single channel conductances of the K^+ channels were reversibly reduced nearly equally from either side of the membrane in a dose dependent manner (*34*). In contrast, mean open dwell time of the channel was elongated with NM (Fig. 2a). These results can reasonably be interpreted in terms of an open channel blocking mechanism where it is assumed that a drug can bind to the blocking site in the open state only. The dose-inhibition curve followed a single site titration curve (Fig. 2b):

$$\gamma/\gamma_0 = \left\{1 + \frac{[\text{AGs}]}{K_b(\text{V})}\right\}^{-1} \tag{2}$$

where γ is the single channel conductance at a given AGs concentration, γ_0 is the single channel conductance in the absence of AGs, and $K_b(\text{V})$ is the blocking constant defined at a given voltage. It is suggested that one AG molecule blocks one K^+-channel of SR in a manner similar to that observed in a lysoPPI channel (*35*).

Several other amino-derivatives showed similar effects of varying severity: polymyxin B > neomycin > kanamycin, streptomycin > ribostamycin > putressin. Interestingly, this order parallels that of ototoxicity and of the inhibitory effect to the above described lysoPPI channel (*35, 37*). The parallelism in the order of several amino derivatives in the three different systems is not trivial because, in the hair cell of the vestibular system, streptomycin is more toxic than neomycin (*10*) while the order is reversed in the inner ear hair cells. This raises several interesting questions: first, "Is an acute AG toxicity caused by a blockade of cation channels of biological membranes?"; second, "Do cation channels of SR and hair cell membranes have similar binding sites for AGs?"; and finally, "Do the binding sites contain polyphosphoinositides?". These questions are now under investigation using an antibody for PPI (*40*).

IV. VOLTAGE DEPENDENT CHANNEL BLOCKADE BY AMINOGLYCOSIDES

The AGs blockade of the SR-K^+ channel was strictly voltage dependent, *i.e.*, single channel conductance at a fixed concentration of AGs gradually decreased with increasing voltage on the same side to which the drug was added. Figure 3a shows a plot of γ/γ_0 *vs.* membrane potential in the presence of 1 mM ribostamycin (RM) on the *cis* side. As seen, γ/γ_0 decreases with membrane potential, while it does not change in the absence of the drug (control). The simplest interpretation of such a voltage dependent channel blockade by ions can be made based on Woodhull's theory of ionic blockade (*7, 41*). The theory assumes that

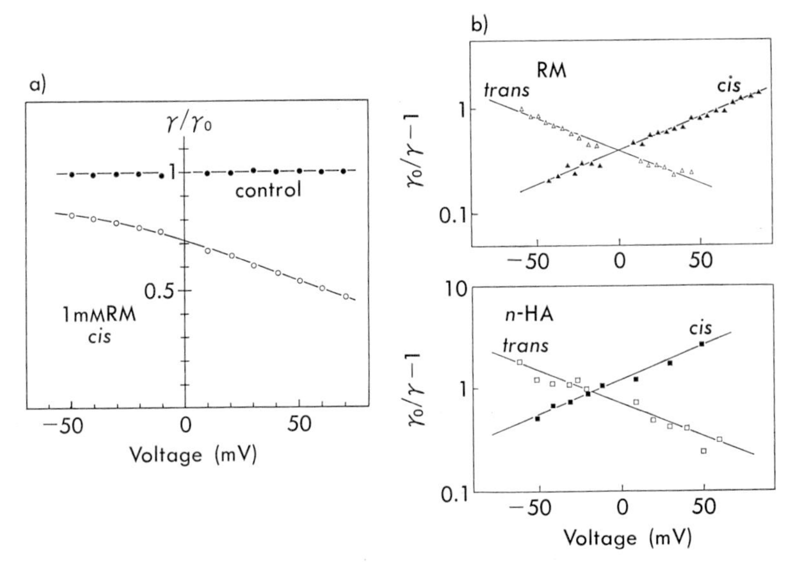

Fig. 3. (a) Voltage dependent single channel conductance in the presence of 1 mM ribostamycin on the *cis* side. Conditions are the same as in Fig. 2. (b) Linearized plot of voltage dependent channel blockade by ribostamycin (RM) and *n*-hexylamine (*n*-HA). Upper trace: 4 mM *cis*-RM (▲) and 3 mM *trans*-RM (△). Lower trace: 1 mM *cis-n*-HA (■) and 0.5 mM *trans-n*-HA (□). The aqueous phase was the same as in Fig. 2a. Blocking parameters determined from these lines are $Z\delta=0.38$ and K_b $(0)=2.6$ mM for *cis*-RM, $Z\delta=-0.36$ and K_b $(0)=1.3$ mM for *trans*-RM, $\delta=0.38$ and K_b $(0)=3.6$ mM for *cis-n*-HA, $\delta=-0.39$ and K_b $(0)=4.4$ mM for *trans-n*-HA. If we assume that the binding sites are the same for both drugs, one amino residue from an AG may bind to the *cis* or *trans* blocking sites.

the blocking ion forms a reversible complex at a site within the electrical field inside the K^+-conducting pathway. In such a case, the electrical field acting on the blocking ion would contribute to the standard free energy of the blocking reaction. Thus, $K_b(V)$, the dissociation constant for the blocking reaction, should vary exponentially with the voltage according to

$$K_b(V)=K_b(0)\exp(-Z\delta FV/RT) \tag{3}$$

where $K_b(0)$ is the zero voltage dissociation constant, Z is the valence of the blocking ion, δ is the fraction of the total electrical potential drop across the membrane found at the blocking site, and F,R,T have their usual meanings. Combining the single channel titration curve (Eq. 2) with Eq. 3, we obtain

$$\gamma/\gamma_0=\left\{1+\frac{[\mathrm{AGs}]}{K_b(0)}\exp(Z\delta FV/RT)\right\}^{-1} \tag{4}$$

This equation is conveniently linearized as follows

$$\ln\left(\frac{\gamma_0}{\gamma}-1\right)=\ln\frac{[\mathrm{AGs}]}{K_b(0)}+Z\delta FV/RT \tag{5}$$

Thus, by measuring single channel conductance as a function of applied voltage at a fixed concentration of AGs, we can test whether the theory can explain the observed voltage dependent channel blocking and, at the same time, get values for $K_b(0)$ and electrical distance, $Z\delta$, from the *cis* side. Figure 3b (upper trace) shows such a linearized plot in a case where RM was present on the *cis* or the *trans* side of the membrane; the plot is nearly linear in both cases. The parameters obtained from this plot are: $Z\delta=0.38$ and $K_b(0)=2.6$ mM for *cis* blocking, and $Z\delta=-0.36$ and $K_b(0)=1.3$ mM for *trans* blocking. If we assume a constant electrical field within the ion conducting pathway, it is suggested that the SR-K^+ channel has two binding sites for AGs which locate nearly symmetrically on the way of the ion conducting pathway.

However, since AGs contain a number of ionizable amino residues, we cannot estimate well the value for Z which contributes to the blocking reaction. So, we used a simpler amino derivative, *n*-hexylamine (*n*-HA), having one ionizable amino residue, *i.e.*, Z is known as 1. The drug blocked the channel in the same way as that for AGs with respect to a single site binding and a voltage dependent blocking from either the *cis* or *trans* side (Fig. 3b, lower trace). Thus, it may be reasonable to use *n*-HA as a model substance for AGs. Interestingly, the values for $Z\delta$ obtained from the linearized plot of voltage dependent blockade by *n*-HA were nearly equal ($Z\delta_{cis}=0.37$, $Z\delta_{trans}=-0.39$) to those from the AGs blockade. Therefore, by assuming that *n*-HA shares the same blocking sites as AGs, we get the value for Z of AGs as 1, suggesting that a certain amino residue from an aminoglycoside binds the blocking site of the SR-K^+ channel. By assuming a constant electrical field across an ion conducting pathway with an estimated length of 10 Å (*25*), it can be concluded that a SR-K^+ channel has two different binding sites located approximately 4 Å distant from the *cis* and *trans* entrances of the channel, respectively, and one amino residue from an AG binds the *cis* or *trans* sites to block the channel.

Basically the same analysis was applied to the macroscopic voltage-dependent conductance inhibition by AGs of the K^+ conductance induced by lysoPPI channels by assuming that the drugs block the channel (*11*). This analysis may be reasonable because the macroscopic K^+ conductance in the absence of AGs and Ca^{2+} was nearly constant within the experimental voltage range (up to ± 120 mV) and because the AGs inhibition of the K^+ conductance of the lysoPPI channel follows the single site titration curve (Fig. 1a). The obtained values for electrical distance, $Z\delta$, of the blocking sites in the lysoPhIP$_2$ and lysoPhIP channels were nearly identical for several AGs (approximately 0.26 and -0.26 for *cis* and *trans* blocking, respectively). In the lysoPPI channel, *n*-HA again inhibited the conductances in the same way as in the AGs inhibition with respect to the single site titration and the voltage-dependent inhibition from either side of the membrane. Thus, we obtained the value for fractional distance δ ($=\pm 0.26$) of the blocking site on the ion-conducting pathway of the lysoPPI channel by the same procedure as used in the SR-K^+ channel (data not shown). It was concluded that a lysoPPI channel has two different AGs-binding sites located symmetrically on the way of the ion conducting pathway, and that a certain amino residue from an AG binds the respective blocking site of the channel as seen in the SR-K^+ channel-blockade by AGs. Fractional distance δ for a Ca^{2+}-blocking site was also determined. The obtained value ($\delta = \pm 0.28$) is nearly identical to that of an AGs binding site indicating that Ca^{2+} and AGs share the same site, as suggested by their competitive relationship (Fig. 1b).

It seems very striking that AGs act on two quite different kinds of channels in a very similar way in several respects. The SR-K^+ channel is equipped with a voltage dependent gate and a relatively strong ion selective filter while the lysoPPI channel lacks both. The most distinct difference between the two channels is their affinity for AGs, the affinity of the SR-K^+ channel being lower than that of the lysoPPI channel by 50–100 fold. However, we recently found that the voltage-dependent gate of the SR-K^+ channel greatly influences the accessibility of AGs (*26*). In a certain state of the gate, the affinity for AGs was increased by several ten-fold approaching the value of the lysoPPI channel. Thus, the AGs-binding site of the SR-K^+ channel seems very similar to that of the lysoPPI with respect to the mode of blocking, affinity, and the blocking order among AGs. Of course, this resemblance may be only a superficial one, but it may be possible that (lyso)PPI participate in part of the function of the ion conducting pathway of the SR-K^+ channel.

SUMMARY

The effects of AGs on the cation channels derived from PPI and SR incorporated in planar lipid bilayers were investigated.

K^+-conductance of the lysoPPI channel was inhibited by the drugs in a dose dependent manner following a single site binding scheme. This inhibitory effect of AGs was, in turn, competitively inhibited by calcium ion which is known to reduce the AGs acute ototoxicity. The order of the conductance inhibition by AGs paralleled that of the AGs ototoxicity. Thus, the lysoPPI channel may be a useful model for a sound transduction channel of hair cells and for understanding the mechanism of the AGs acute ototoxicity occurring on the cation channels of a hair cell in the inner ear.

To learn the detailed modes of action of AGs on biological channels, AGs effects on the single channel behavior of a cation channel from SR membranes were investigated. It was clearly demonstrated that AGs behave as open channel blockers to the SR channel; the modes of action (dose- and voltage-dependency) were the same as those observed in the lysoPPI channel. The order of the channel blocking effect among AGs again paralleled that of the AGs acute ototoxicity.

The common AGs effect among three different systems (lysoPPI-channel, SR-cation channel and perhaps the transduction channel of a hair cell) suggests that the lysoPPI channel may not only be a useful model for certain biological channels, but also actually have a role in ion transport across biological membranes.

Acknowledgments

The author heartily acknowledges his coworkers Mr. J. Hayase and Mr. Y. Oosawa. Thanks are also due to Prof. K. Miyamoto, Prof. M. Kasai, Dr. T. Yoshioka, and Dr. F. Hayashi for their fruitful discussions and encouragement. This work was supported in part by the Sakkokai Foundation and grant No. 58780239 from the Ministry of Education,

Science and Culture of Japan. Some parts of this article will be published elsewhere as original papers.

REFERENCES

1 Abdel-Latif, A.A., Green, K.K., Smith, J.P., McPherson, Jr., J.C., and Matheny, J.L. (1978). *J. Neurochem.* **30**, 517–525.

2 Akiyoshi, M., Yano, S., Shoji, T., Tajima, T., Inazawa, T., Hara, T., and Shimizu, M. (1977). *Chemotherapy* **25**, 1892–1914.

3 André, T. (1956). *Acta Radiol.* Suppl. **142**, 1–89.

4 Birnberger, A.C., Birnberger, K.L., Eliasson, S.C., and Simpson, P.C. (1971). *J. Neurochem.* **18**, 1291–1298.

5 Brockerhoff, H. and Ballou, C.E. (1962). *J. Biol. Chem.* **18**, 1291–1298.

6 Corey, D.P. and Hudspeth, A.J. (1979). *Nature* **281**, 675–677.

7 Coronado, R. and Miller, C. (1979). *Nature* **280**, 807–810.

8 Dawson, R.M.C. (1969). *Ann. N.Y. Acad. Sci.* **165**, 774–783.

9 Gerrad, J.M., Kindom, S.E., Peterson, D.A., Krantz, J.P.K.E., and White, J.G. (1979). *Am. J. Pathol.* **96**, 423–436.

10 Hawkins, J.E., Jr. (1976). In *Handbook of Sensory Physiology; Auditory System*, ed. Keidel, W.D. and Neff, W.D., pp. 707–748. Berlin, Heidelberg: Springer-Verlag.

11 Hayase, J. and Sokabe, M. (1983). *Biophysics*, **23**(Suppl.), S53.

12 Hayashi, F., Sokabe, M., Hayashi, K., Takagi, M., and Kishimoto, U. (1978). *Biochim. Biophys. Acta* **510**, 305–315.

13 Hayashi, F., Sokabe, M., and Amakawa, T. (1981). *Proc. Japan Acad.* **57B**, 48–53.

14 Hendrickson, H.S. (1969). *Ann. N.Y. Acad. Sci.* **165**, 668–676.

15 Hendrickson, H.S. and Reinertsen, J.L. (1971). *Biochem. Biophys. Res. Commun.* **44**, 1258–1264.

16 Hladky, S.B. and Haydon, D.A. (1972). *Biochim. Biophys. Acta* **274**, 294–312.

17 Hudspeth, A.J. (1982). *J. Neurosci.* **2**, 1–10.

18 Hudspeth, A.J. and Corey, D.P. (1977). *Proc. Natl. Acad. Sci. U.S.* **74**, 2407–2411.

19 Kai, M. and Hawthorne, J.H. (1969). *Ann. N.Y. Acad. Sci.* **165**, 761–773.

20 Lattore, R. and Benos, D. (1984). This volume, pp. 199–213.

21 Läuger, P. (1972). *Science* **178**, 24–30.

22 Michell, R.H. (1975). *Biochim. Biophys. Acta* **415**, 81–147.

23 Michell, R.H. (1979). *Trends Biochem. Sci.* **4**, 128–131.

24 Miller, C. (1978). *J. Memb. Biol.* **40**, 1–23.

25 Miller, C. (1982). *J. Gen. Physiol.* **79**, 869–891.

26 Putney, J.W. Jr., Weiss, S.J., Van de Walle, C.M., and Haddas, R.A. (1980). *Nature* **284**, 345–347.

27 Robson, J.M. and Sullivan, F.M. (1963). *Pharmacol. Rev.* **15**, 169–223.

28 Sanders, W.E. Jr. and Sanders, C.C. (1979). *Annu. Rev. Pharmacol. Toxicol.* **19**, 53–83.

29 Schacht, J. (1976). *J. Acoust. Soc. Am.* **59**, 940–944.

30 Schacht, J. (1984). This volume, pp. 89–97.

31 Schacht, J., Weiner, N.D., and Lodhi, S. (1978). In *Cyclitols and Phosphoinositides*, ed. Wells, W.W. and Eisenberg, F., pp. 153–165. New York: Academic Press.

32 Serhan, C.N., Fridovich, J., Goetzl, E.J., Dunham, P.B., and Weismann, G. (1982). *J. Biol. Chem.* **257**, 4746–4752.
33 Sokabe, M. (1982). *Bull. Fac. Human Sci. Osaka Univ.* **8**, 241–288.
34 Sokabe, M. (1983). *Proc. Japan Acad.* **59B**, 33–37.
35 Sokabe, M., Hayase, J., and Miyamoto, K. (1982). *ibid.* **58B**, 177–180.
36 Sokabe, M., Oosawa, Y., and Kasai, M. (1983). *Proc. Jpn. Soc. Gen. Com. Physiol.* 40.
37 Tachibana, M. (1984). This volume, pp. 101–115.
38 Torda, C. (1974). *Int. Rev. Neurobiol.* **16**, 1–66.
39 Tret'jak, A.G., Limarenko, I.M., Kossava, G.V., GuLak, P.V., and Kozlov, Yu.P. (1977). *J. Neurochem.* **28**, 199–205.
40 Tsukamoto, T., Yoshioka, T., and Sokabe, M. (1983). *Biophysics* **23** (Suppl.), S54.
41 Woodhull, A.M. (1973). *J. Gen. Physiol.* **61**, 687–708.

11

THERMOSENSORY TRANSDUCTION IN *PARAMECIUM*

YASUO NAKAOKA

Department of Biophysical Engineering, Osaka University, Osaka 560, Japan

Many species of organisms have thermal sensitivity and detect a sudden temperature change. In multicellular organisms there are specialized thermosensory cells (*1*), and in unicellular organisms the cell itself is the sensory cell. The mechanism of how cells detect temperature change is not known.

A ciliated protozoon, *Paramecium*, has a high sensitivity to temperature change and shows thermotaxis to accumulate at the cultured temperature in a temperature gradient (*2*). The electrophysiology of the paramecium cell is also advanced (*3*). These factors stimulated us to investigate the temperature sensitive behavior and to measure the membrane potential changes induced by sudden temperature change.

We also investigated the change in membrane potential during adaptation to the K^+ concentration of the medium. Membrane potential initially hyperpolarized or depolarized by the transfer to a lower or higher concentration of K^+, recovered nearly the original value within 1–2 hr.

Experimental analyses have shown that these responses in the membrane potential are caused by the changes in the K^+ conductance of the membrane.

I. TEMPERATURE SENSITIVE-BEHAVIOR

When paramecium cells cultured at 25°C were placed in a temperature-controlled vessel and the temperature was suddenly dropped from 25 to 20°C, the cells transiently increased the frequency of their directional changes in swimming, then decreased and within 1 min approached the stationary value (Fig. 1a). When the temperature was inversely changed from 20 to 25°C, the frequency transiently decreased, and within 1 min returned to the initial value (Fig. 1b).

The magnitude of the transient response in the frequency of directional change depended on the rate of temperature change (*4*).

The maximum transient response appeared 5–10 sec after onset of

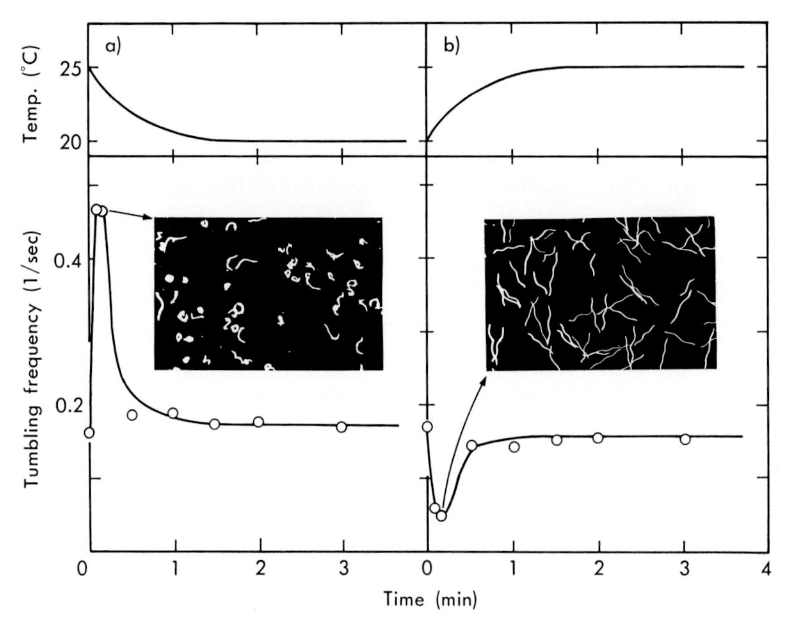

Fig. 1. Response to temperature change. The tumbling frequency was measured on tracks in photographs taken at suitable intervals after the beginning of temperature change. Photographs were taken at 2 sec exposure at the time indicated by arrows. The temperature measured by a thermistor is shown in the upper part of the figure.

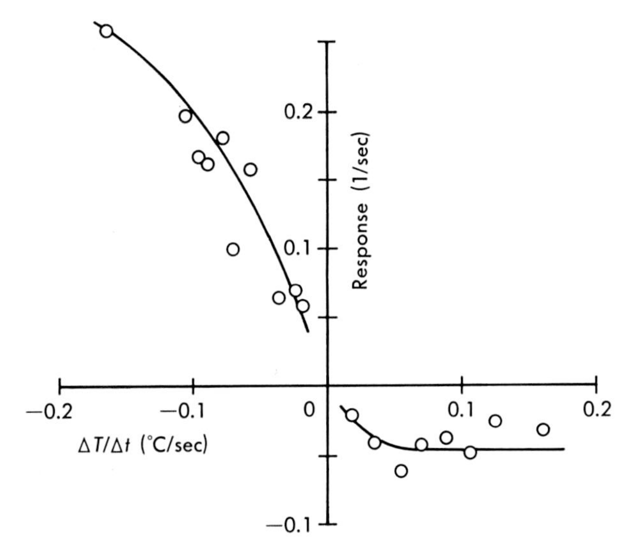

Fig. 2. Relation between the transient response in tumbling frequency and the rate of temperature change.

the temperature change, and therefore, this maximum was determined as a function of the rate of temperature change (Fig. 2). The relation shows that even at a very slow temperature drop, *e.g.*, 0.015°C/sec, a transient increase in the frequency is induced and with an increasing rate of change, the magnitude of the response increases.

In a temperature gradient, the cells sense the spatial gradient of temperature as the temporal change of temperature during random swimming (*4*, *7*). Actually, when cells cultured at 25°C were put in a temperature gradient vessel whose temperature at the respective ends was 20 and 25°C, the frequency of directional changes of cells descending the temperature gradient was much greater than that of the cells ascending. As the result of these differences in the frequency of directional changes, the cells accumulated at 25°C.

II. THERMO-RECEPTOR POTENTIAL (*8*)

A paramecium cell cultured at 25°C was impaled with two intracellular electrodes, one to inject current and one to record membrane potential, and the temperature of the surrounding medium was suddenly changed. Following the temperature drop from 25 to 20°C, the membrane potential slowly depolarized and repolarized to the original level (Fig. 3). At the initial phase of depolarization, some spike-like depolarizations were recorded. On the other hand, following a temperature rise from

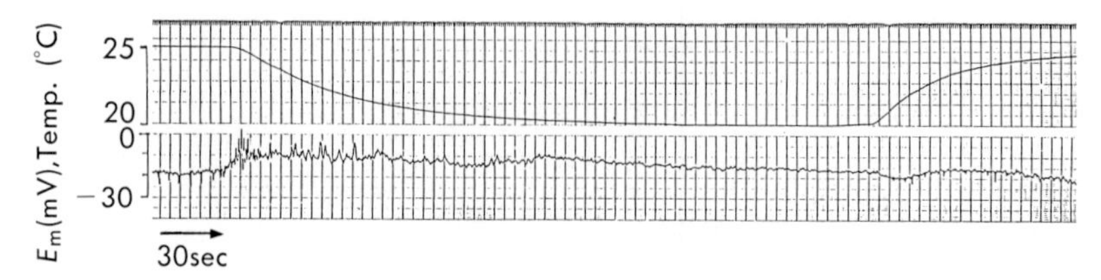

Fig. 3. Transient changes in membrane potential induced by sudden temperature changes. Lower record shows the membrane potential of *Paramecium* cultured at 25°C. Upper record is the temperature of the surrounding medium.

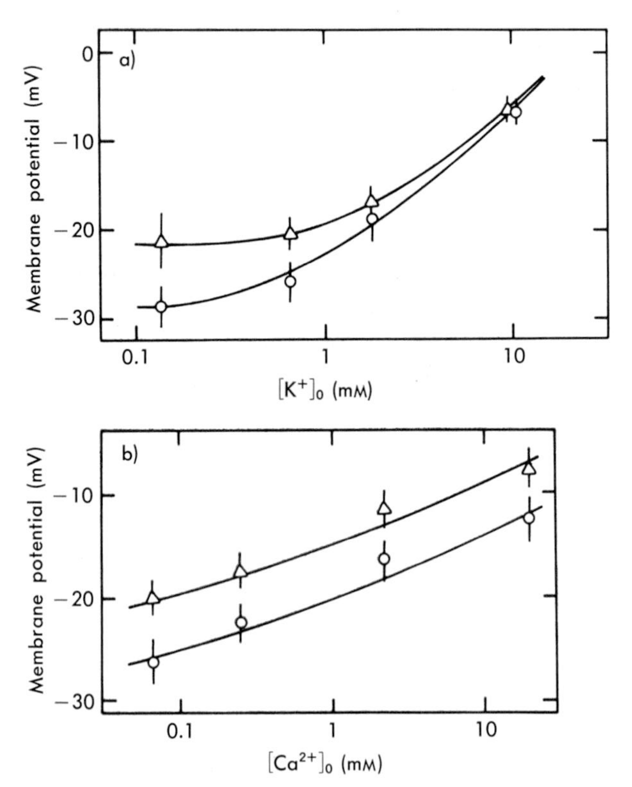

Fig. 4. Dependence of receptor potentials upon external concentrations of K^+ (a) and Ca^{2+} (b). Changes in the membrane potential of cells cultured at 25°C were measured following the temperature drop from 25 to 20°C. Resting potential (○) and peak value of the depolarizing receptor potential (△) are plotted.

20 to 25°C, the potential slowly hyperpolarized during which spike frequency decreased, then recovered to the original value. As the depolarizing spike is considered to induce the directional change in the swimming, the increase or decrease in spike frequency during the slow potential changes correspond to the respective increase or decrease in the frequency of such directional change in the swimming. These slow potential changes are thermoreceptor potentials induced by sudden temperature changes.

To know the ionic mechanism generating the thermo-receptor potential, the dependence of the receptor potential on Ca^{2+} or K^+ concentration of external medium was examined. With increasing K^+ concentration, the amplitude of depolarizing receptor potential decreased (Fig. 4a). The slope of peak value of receptor potential versus K^+ concentration was smaller than that of the resting potential. Increasing the Ca^{2+} concentration, however, did not change the amplitude (Fig. 4b). As the slope of membrane potential versus external ion concentration is a measure of membrane conductance of the ion, this result suggests that the membrane conductance of K^+ decreases with the depolarizing receptor potential.

When the conductance of the cell membrane was successively measured by injecting a small hyperpolarizing current, at the initial phase of depolarizing receptor potential the conductance was slightly decreased, and at the initial phase of hyperpolarizing receptor potential the conductance was increased. As the membrane conductance of a paramecium cell is mainly composed of the K^+ conductance (*3*), these changes in conductance following temperature change are probably caused by the change in the membrane conductance of K^+.

III. K^+-SENSITIVE BEHAVIOR (*5*)

When paramecium cells were transferred to a solution with a lower concentration of K^+ than one to which they had previously adapted, they initially accelerated forward swimming by 2.5–3 times, then gradually reduced it to a stationary value (Fig. 5). On the other hand, when the cells were transferred to a solution of a higher K^+ concentration, they showed backward swimming or circling for a while, then returned to a normal mode. Within a limited K^+ concentration range,

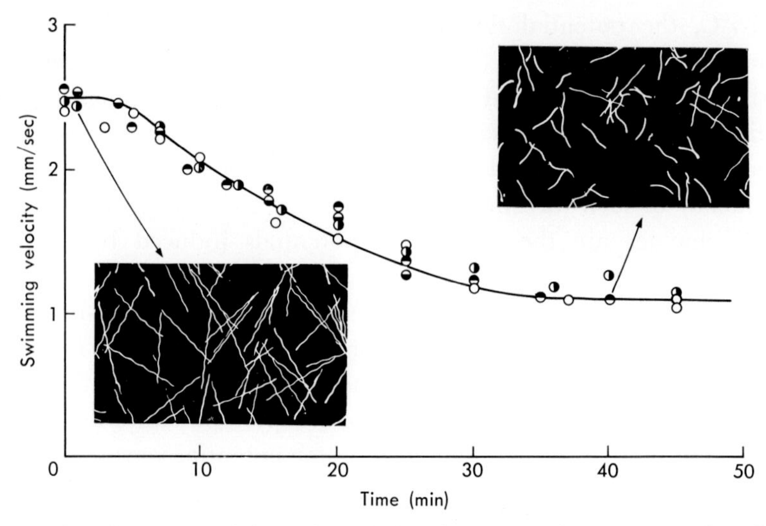

Fig. 5. Time course of decreasing velocity after acceleration. *Paramecia* adapted to solutions of 0.25 mM $CaCl_2$ and 2.5 mM (◓), 5 mM (◑), 7.5 mM (◒) or 10 mM (○) KCl were transferred to a solution of 0.25 mM $CaCl_2$ and 1 mM KCl, after which swimming velocity was measured. Photographs were taken at 2 sec exposure at the time indicated by arrows.

the cells recovered the stationary swimming almost unaffected by the ionic conditions.

IV. CHANGE IN MEMBRANE POTENTIAL (*6*)

A close relationship between the membrane potential and the ciliary beating has been shown in *Paramecium* (*3*). Hyperpolarization of the potential causes an increase in beat frequency of cilia, and depolarization causes a ciliary reversal. As an increase or a decrease in K^+ concentration of surrounding medium respectively depolarizes or hyperpolarizes the membrane potential, the swimming behavior is initially changed by a transfer to a solution of different K^+ concentration. In adapting to the new solution, the cells recover normal swimming. The cells are expected to change their membrane potential during adaptation.

Figure 6 shows that the dependence of potential upon the external K^+ concentration shifts to the depolarizing direction as the cell adapts to decreasing K^+ concentration. These shifts of the potential result in maintaining it at nearly a constant value in the adapted solution. That

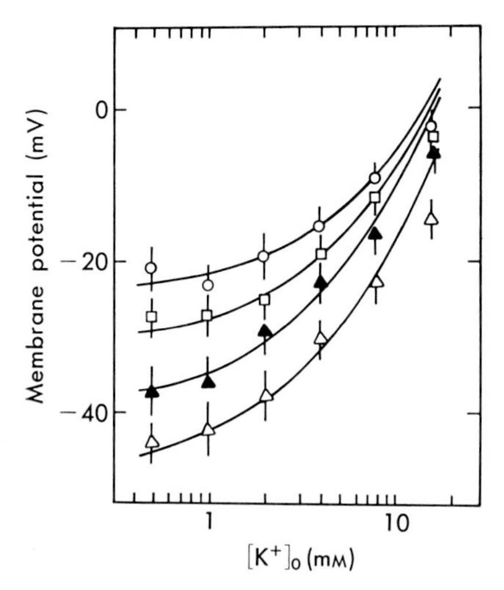

Fig. 6. Effect of $[K]_0$ on the membrane potential of cells adapted to various concentrations of K^+. The cells were adapted to solutions containing 0.25 mM $CaCl_2$ and 2 mM (○), 4 mM (□), 8 mM (▲) or 16 mM (△) KCl, and membrane potentials were measured in the solution containing 0.25 mM $CaCl_2$ and KCl indicated in the abscissa. Arrows indicate the potentials of the condition in which the cells have been adapted. The curves were drawn to fit measured points using the following equation with $K^+_i = 20$ mM (○,□), 22 mM (▲) and 27 mM (△), and $P_{Ca}/P_K = 11$ (○), 8 (□), 6 (▲), and 4 (△).

$$E_m = \frac{RT}{F} \ln \frac{P_K[K^+]_0 - P_K[K^+]_i + \sqrt{A}}{2P_K[K^+]_i + 8P_{Ca}[Ca^{2+}]_i}$$

where, $A = P_K^2([K^+]_i - [K^+]_0)^2 + 4(P_K[K^+]_0 + 4P_{Ca}[Ca^{2+}]_0)(P_K[K^+]_i + 4P_{Ca}[Ca^{2+}]_i)$

is, the potential of cells transferred to a lower concentration of K^+ initially hyperpolarized, then depolarized gradually as they adapted to the solution. Inversely, the potential of cells transferred to a higher concentration of K^+ initially depolarized, then gradually hyperpolarized. Finally, the potential largely recovered that before the transfer and was within −15 to −20 mV in the adapted solution.

Accompanying the adaptation to different K^+ concentrations, the membrane resistance measured by the injection of a small hyperpolarizing current was changed. The conductance of cells adapted to 8 mM K^+ was 1.5–1.7 times as large as that of cells adapted to 2 mM K^+. On the other hand, an intracellular concentration of K^+ of the cells adapted to 8 mM K^+ was 2–3 mM higher than those adapted to 2 mM K^+. Therefore, the change in membrane potential during adaptation is mainly caused by the modulation of K^+ conductance. How the paramecium cells regulate this conductance during the adaptation is the next problem to be studied.

SUMMARY

A ciliated protozoon, *Paramecium*, senses a temporal change in environmental conditions, *e.g.*, temperature and ionic concentrations, and this is expressed in the swimming behavior. With a sudden change from the culture temperature to one lower or higher, paramecium cells transiently increase the frequency of directional changes in swimming. Following a drop from the culture temperature, a transient, slow depolarization of the potential is induced during which depolarizing spikes are frequently generated to induce the reversal of ciliary beating. When paramecium cells are transferred to solution with a lower concentration of K^+, their swimming velocity increases by 2.5–3 times as a result of the hyperpolarization of the membrane potential. During adaptation to the solution, the swimming velocity decreases and the membrane potential recovers to nearly the original level. These responses in the membrane potential are induced mainly by the modulation of membrane conductance of K^+.

REFERENCES

1 Hensel, H. (1973). In *Handbook of Sensory Physiology*, vol. 2, ed. Iggo, A., pp. 79–110. Heidelberg, New York: Springer.
2 Mendelssohn, M. (1895). *Pflüger's Arch. Ges. Physiol.* **60**, 1–27.
3 Naitoh, Y. and Eckert, R. (1974). In *Cilia and Flagella*, ed. Sleigh, M.A., pp. 305–352. London, New York: Academic Press.
4 Nakaoka, Y. and Oosawa, F. (1977). *J. Protozool.* **24**, 575–580.
5 Nakaoka, Y., Oka, T., Serizawa, K., Toyotama, H., and Oosawa, F. (1983). *Cell Struct. Funct.* **8**, 77–84.
6 Oka, T. (1983). Changes of calcium and potassium channels during adaptation to ionic solution in *Paramecium*. Ph.D. thesis, Osaka University.
7 Tawada, K. and Miyamoto, H. (1973). *J. Protozool.* **20**, 289–292.
8 Toyotama, H. (1981). Thermo-receptor potential in *Paramecium*. Ph.D. thesis, Osaka University.

12

GENETICS OF SENSORY TRANSDUCERS IN *ESCHERICHIA COLI*

JOHN S. PARKINSON, ANN M. CALLAHAN, AND
MARY K. SLOCUM

Biology Department, University of Utah, Utah 84112, U.S.A.

Motile bacteria exhibit locomotor responses to a variety of environmental stimuli. *Escherichia coli*, for example, is attracted to sugars and amino acids that can serve as carbon and energy sources, and repelled by potentially harmful agents such as alcohols, fatty acids, indole, and certain divalent metal cations. In addition, these organisms respond to gradients of oxygen and temperature. In order to carry out behavior of this sort, the cells must be able to monitor and sense changes in their environment, to transduce sensory information into signals that regulate motor activity, and to integrate multiple or conflicting stimuli. Due to its relative simplicity and its suitability for both genetic and biochemical studies, the chemotaxis system of *E. coli* is an excellent model for exploring stimulus transduction processes at the molecular level.

I. CHEMOTACTIC BEHAVIOR OF *E. COLI*

In uniform chemical environments, bacteria swim about in a three-dimensional random walk consisting of smooth swimming punctuated by tumbling episodes which reorient the cells. Both smooth and tumbly movements are produced by rotation of flagellar filaments, the locomotor organelles. Rotation in the counter-clockwise (CCW) direction causes smooth swimming, whereas clockwise (CW) rotation produces tumbling. Chemotactic migration in chemical gradients is brought about by altering the pattern of flagellar rotation in response to temporal changes in attractant or repellent concentration (*1*, *7*). For example, when a cell happens to head toward an attractant source, it senses an increase in attractant concentration with time, which in turn reduces the likelihood of CW rotation and tends to prolong swimming in the favorable direction. Experiments in which individual cells are subjected to temporal changes in chemoeffector concentration reveal two phases to the chemotactic responses: a rapid excitation phase during which the stimulus is detected and a change in flagellar rotation is initiated; and a slower adaptation phase which culminates in a return to prestimulus swimming behavior even though the stimulating chemical is still present. The adaptation process enables the cell to detect temporal changes in chemical concentration as it swims about in spatial gradients.

II. THE CHEMOTAXIS MACHINERY

Approximately 50 different gene products are required for motility or chemotaxis in *E. coli*. This chemotaxis machinery comprises a network of signaling elements through which sensory information about the chemical environment is transmitted to the flagella (*12*). Several integral membrane proteins termed "methyl-accepting chemotaxis proteins" or MCP's play important roles in both the excitation and adaptation phases of chemotaxis (*17*). The MCP molecules receive information about the chemical environment either directly through specific binding of small molecules, or through interaction with ligand-occupied binding proteins in the periplasmic space. Upon detecting a change in attractant or repellent concentration, the MCP's generate or modulate signals that control the direction of flagellar rotation, thereby eliciting a locomotor response. Finally, the MCP's undergo covalent modifications that alter their signaling properties and lead to sensory adaptation. The molecular details of the excitation and adaptation processes mediated by MCP molecules are still poorly understood. Nor do we know the nature of the flagella-controlling signal(s) generated by the transducer proteins. In the remainder of this article we discuss some recent work on the structure and function of bacterial transducers which is beginning to shed light on these questions.

III. TRANSDUCER GENES AND PROTEINS

Four different MCP structural genes have been identified in *E. coli* (Table I). Each synthesizes an integral membrane protein of approximately 60,000 daltons in size. The chemotactic responses mediated by these transducers have been examined by isolating and characterizing mutants with a complete loss of activity due to deletion, insertion or nonsense mutations (*2*, *10*). The *tsr* product is required for chemotactic responses to serine, a strong attractant, and has been shown to bind this compound with high affinity (*3*). Other behavioral responses such as weak acid repulsion and thermotaxis are also altered in *tsr* mutants; however, it is not yet clear whether the *tsr* product plays a direct or indirect role in these events. The *tar* gene product is required for responses to the attractants aspartate and maltose, and is also involved in responses to divalent metal cation repellents. The *tar* protein detects aspartate through a high affinity binding site (*19*), but senses maltose stimuli by interacting with a periplasmic maltose binding protein, which

TABLE I
E. coli Genes Involved in Sensory Transduction and Adaptation.

		Null phenotype[a]					
		Flagellar		Attractant responses[c]			
Gene	Product	rotation[b]	Ser	Asp	Mal	Rib	Gal
tsr	MCPI	Wild type	−	+	+	+	+
tar	MCPII	Wild type	+	−	−	+	+
trg	MCPIII	Wild type	+	+	+	−	−
tap	MCPIV	Wild type	+	+	+	+	+
cheR	Methyltransferase	CCW bias		Respond, but do not adapt			
cheB	Methylesterase	CW bias		Respond, but do not adapt			

[a] The null phenotype is defined by mutations such as deletions that have no detectable gene product or functional activity.

[b] These are the unstimulated patterns (wild type=no significant difference from wild type behavior). Cells with rotational biases typically have very low reversal rates, with most of the cells rotating exclusively in one direction.

[c] Attractants: Ser, serine; Asp, aspartate; Mal, maltose; Rib, ribose; Gal, galactose.

serves as the maltose chemoreceptor (*5*). The *trg* product is needed for both ribose and galactose responses. It does not appear to bind these attractants directly, but rather probably interacts with periplasmic binding proteins for these two sugars.

Mutants with deletions of the *tap* locus, the fourth MCP gene of *E. coli*, exhibit wild type responses to all known chemotactic stimuli (*10*). Nucleotide sequence comparisons of the various MCP genes indicate that the *tap* locus contains highly conserved regions characteristic of all transducer molecules (*6*). This finding argues that Tap structure has been maintained by selection, but thus far no transducer activity has been demonstrated for this MCP.

IV. COVALENT MODIFICATIONS OF TRANSDUCERS

When the cell detects and responds to a chemotactic stimulus, the methylation state of transducer molecules engaged in excitatory signaling changes: favorable (CCW-enhancing) stimuli elicit increases in methylation, whereas unfavorable (CW-enhancing) stimuli cause decreases in methylation (*16*). Flagellar responses cease when a new steady-state methylation level is reached, suggesting that changes in transducer methylation are involved in the process of sensory adaptation. The methylation state of transducer molecules is determined by the relative activities of two cytoplasmic enzymes, the *cheR* gene product, which is an MCP-specific methyltransferase (*16*), and the *cheB* gene product, which is an MCP-specific methylesterase (*18*). The *cheR* enzyme catalyzes transfer of methyl groups from *S*-adenosylmethionine to the gamma carboxyl of several glutamic acid residues in MCP molecules. The *cheB* enzyme hydrolyzes these methyl esters, yielding methanol and a restored glutamate moiety. The behavior of *cheR* and *cheB* deletion mutants, which totally lack the ability to add or remove MCP methyl groups, is summarized in Table I. Mutants defective in *cheR* function exhibit an extreme CCW bias in flagellar rotation, but can still respond to strong CW-enhancing stimuli such as repellent increases (*9*). Conversely, *cheB* mutants exhibit a severe CW bias, but still respond to CCW-enhancing stimuli such as attractant increases (*20*). Unlike wild type cells, however, *cheR* and *cheB* mutants do not adapt to stimuli, but rather continue to display a flagellar response as long as the stimulating chemical is present. These findings clearly indicate that transducer methylation-demethylation is not required for the excitation phase of a chemotactic response, but is required for sensory adaptation.

The aberrant rotational patterns of *cheR* and *cheB* mutants suggest that methylation and demethylation affect the signaling properties of transducer molecules: addition of methyl groups favors CW rotation, whereas removal of methyl groups favors CCW rotation. Since stimulus-induced changes in transducer methylation patterns are confined to the transducer that handles the stimulus, it would seem that excitation induces a change in the substrate properties of the transducer molecules which leads to net addition or removal of methyl groups and cancellation of the excitatory signal. In addition, the activities of the *cheR* and *cheB* gene products may be regulated by a feedback loop from the flagellar switching mechanism (*12*).

A second post-translational modification of transducer molecules has been demonstrated recently, although its physiological significance is not yet clear (*13*, *14*). Several of the methylation sites in mature MCP molecules are actually glutamine residues which have been deamidated during or after MCP synthesis (*4*). This deamidation reaction is dependent on the presence of functional *cheB* product, and may be catalyzed by this enzyme since glutaminases often have esterase activity. The chemical consequences of deamidation are similar to those of demethylation (generation of new negative charges on MCP molecules) and have similar effects on the signaling properties of transducers. Deamidated or unmethylated molecules are correlated with CCW flagellar rotation, whereas undeamidated or methylated molecules are correlated with CW rotation. The undeamidated MCP molecules present in *cheB* mutants can still accept a few methyl groups, and function with near normal efficiency in excitation, so deamidation is probably not essential for transducer activity. Because deamidation is irreversible, whereas methylation-demethylation is not, it has been suggested that the deamidation reaction was a forerunner of the more efficient methylation reactions used for sensory adaptation today (*8*).

V. TRANSDUCER STRUCTURE

The primary structures of the *E. coli* transducer proteins have been

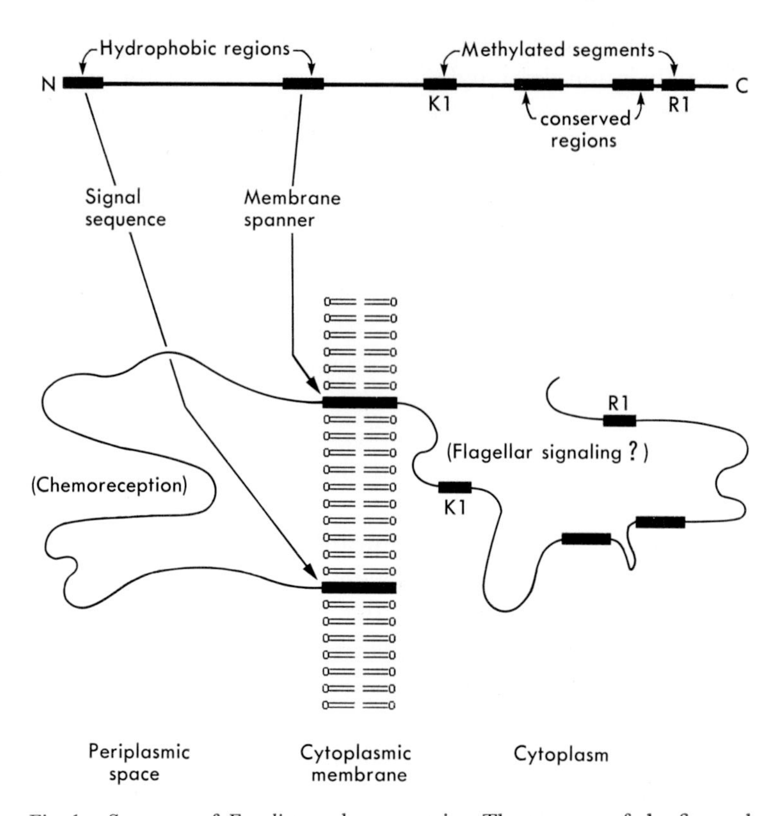

Fig. 1. Structure of *E. coli* transducer proteins. The top part of the figure shows characteristic features of MCP structure revealed by nucleotide sequence analyses of MCP genes (*6*). The figure is drawn roughly to scale; the proteins are approximately 550 amino acid residues in length. The lower part of the figure presents a working model of structure-function relationships in these transducer molecules. The various functional activities appear to be encoded in separate domains of the molecule, which spans the cytoplasmic membrane.

deduced from the nucleotide sequences of the *tar*, *tsr*, *trg* and *tap* genes (*6*; G. Hazelbauer, personal communication). The principal conclusions of that work are summarized in Fig. 1. Each transducer contains two regions of hydrophobic residues in the amino half of the molecule. The first region occurs at the very amino terminus and is believed to represent a signal sequence that initiates transfer of the nascent protein into the cytoplasmic membrane. It is not known whether this leader sequence is subsequently removed or whether it remains in the mature molecule. The second hydrophobic region occurs about 200 residues from the amino end and is thought to be a stop-transfer sequence that serves to anchor the protein in the cytoplasmic membrane. Thus the amino-terminal portion of the transducers probably protrudes into the periplasmic space, whereas the carboxy-terminal portion remains in the cytoplasm. The amino termini of different transducers are quite variable in sequence, suggesting that these regions have different functional activities in different transducers. It seems likely that the receptor functions of transducers are encoded in this portion of the molecule. The carboxyl half of the transducer molecules has two segments that contain methylated glutamyl residues, termed K1 and R1. These segments and the region between them are highly conserved in all transducers, suggesting that this portion of the transducer molecules may contain domains that interact with the CheR and CheB proteins or with other common elements of the chemotaxis system.

It appears from these structural studies that the different functional activities of bacterial transducers may reside in discrete domains of the transducer molecules. For example, the binding sites for chemoeffector compounds and for periplasmic binding proteins are expected to be in the amino-terminal portion of the molecule, whereas the sites for interaction with the methylation and demethylation enzymes should be in the carboxy-terminal portion. However, the current information about transducer structure provides few clues as to which portions of the molecule might be involved in flagellar signaling. We are using genetic methods to analyze structure-function relationships in the *tar* and *tsr* gene products in an attempt to test the domain hypothesis and to identify the transducer properties critical for signaling activity. Below we describe some of the transducer defects we have obtained and the information they provide about the functional architecture of these molecules.

VI. ISOLATION AND CHARACTERIZATION OF TRANSDUCER MUTANTS

Double mutants lacking both *tar* and *tsr* function exhibit an extreme CCW bias in flagellar rotation and are generally nonchemotactic. Although the basis for this effect is not understood, it offered a simple method for recognizing new mutants defective in either of these MCP functions. We constructed specialized lambda transducing phages carrying the *tar* or *tsr* loci, subjected them to mutagenesis, and looked for mutant phage that were unable to correct the generally nonchemotactic phenotype of *tar tsr* strains by complementation. This procedure yielded a variety of *tar* and *tsr* mutations affecting different transducer functions, including many which appeared to be defective in signaling activity or in interactions with heterologous MCP's.

Each of these new mutations was examined for response to nonsense suppressors, for polarity properties, and for dominance effects. The relative physical position of each mutation within the transducer coding region was determined by fine structure deletion mapping, by restriction endonuclease mapping of transducing phages and by marker rescue studies using *tar* and *tsr* fragments cloned into plasmid vectors. We are now in the process of determining the nucleotide sequence changes in each mutation in order to establish the nature of the alterations in the mutant transducer molecules. Phages bearing transducer mutations were used to program protein synthesis in infected host cells. The mutant proteins were then examined by SDS-PAGE analysis to determine whether they were stable, altered in size or mobility, and capable of undergoing post-translational processing. Finally, we examined the unstimulated pattern of flagellar rotation in the transducer mutants and their behavioral responses to spatial and temporal gradients of various compounds. On the basis of these tests, we were able to divide the transducer mutants into a number of discrete classes (Table

TABLE II
Properties of *tar* and *tsr* Mutations

Class	Gene	No. of isolates	Chemotactic behavior	Flagellar rotation	Comments[a]
"Cryptic"	*tar*	1	Wild Type	Wild Type	Che$^-$ in *tsr*$^-$ background
	tsr	3	Wild Type	Wild Type	Che$^-$ in *tar*$^-$ background
"Null"	*tar*	11	(Tar$^-$)	Wild Type	(Full) loss of function
	tsr	31	(Tsr$^-$)	Wild Type	(Full) loss of function
"Tumbly"	*tar*	1	Tar$^{\pm}$	CW-biased	Leaky; aberrant product
	tsr	6	(Tsr$^-$)	CW-biased	Variable taxis defects
"Smooth"	*tar*	13	(Che$^-$)	CCW-biased	(Partly) polar mutations
	tsr	14	(Che$^-$)	CCW-biased	Dominant *cheD* mutations

[a] Brackets indicate that members of a particular class were not uniform with respect to that property.

II). Below we discuss a few of these mutant types and the clues they have provided about transducer structure-function relationships.

VII. TRANSDUCER MUTANTS WITH CRYPTIC DEFECTS

One of the new *tar* mutations exhibited wild type Tar function in a *tsr*$^+$ background, but no Tar function in a *tsr*$^-$ background. Three of the *tsr* mutations displayed analogous behavior with respect to the presence or absence of a functional *tar* locus in the genetic background. These "cryptic" mutations suggest that the *tsr* and *tar* gene products may interact with one another. In wild type cells, such interactions are probably not required for transducer function, because Tsr function is normal in *tar* mutants and *vice versa*. Evidently the cryptic mutations alter the transducers in such a manner that they require the presence of heterologous, unaltered transducers in order to function. The nature of these transducer alterations is not yet known. It could be that certain structural defects in MCP molecules can be suppressed by formation of multimeric complexes with normal transducer subunits. Alternatively, it is conceivable that the cryptic mutations destabilize the transducer molecule, perhaps rendering it susceptible to proteolytic attack, and that interaction with a normal subunit overcomes this defect. In any event, the cryptic phenotype appears to be caused by a specific type of transducer structural defect because all of the cryptic mutations obtained thus far affect the extreme carboxyl end of the molecule.

VIII. TRANSDUCER MUTANTS WITH PROCESSING DEFECTS

The transducer mutants with complete loss of function exhibited considerable variability in the properties of their mutant gene products. Many of these mutants either synthesized no detectable product or made a grossly aberrant product that was shorter than normal or exhibited intrinsic mobility shifts due to amino acid substitution. Such proteins were incapable of being deamidated or methylated. A number of these mutations were mapped to the extreme amino-terminal coding region of the transducer genes, and may affect the signal sequence needed for insertion and transfer of the nascent transducer polypeptide through the cytoplasmic membrane. If transducer export were blocked, it seems likely that the protein would assume an aberrant conformation that might affect the carboxyl as well as the amino domains, thus preventing normal post-translational processing. A few mutants with null behavioral phenotypes made products that could be deamidated but not methylated. This result was quite unexpected, because deamidation and methylation take place in the same region of the transducer molecules, even at the same sites in some cases (*4*). Even more surprising is the fact that these mutations are located in the amino-terminal coding portion of the gene, and are clearly outside of the hydrophobic leader sequence. If our working model is valid, these findings imply that alterations in the receptor domain can influence interactions between the CheR or CheB enzymes and their substrate sites in the carboxyl end of the transducer molecules. Perhaps transducer molecules must normally be in an excited state in order to accept methyl groups, but not to undergo deamidation, and these mutant transducers cannot be excited due to chemoreception defects.

IX. TRANSDUCER MUTANTS WITH SIGNALING DEFECTS

Complete loss of *tar* or *tsr* function has no effect on the unstimulated swimming behavior of *E. coli* (see Table I); however, some of our new mutants exhibited aberrant swimming patterns due to a CCW or CW bias in flagellar rotation. The most dramatic examples are the *cheD* alleles of *tsr*, which cause nearly complete loss of CW rotation and a generally nonchemotactic phenotype (*11*). These mutants are dominant in complementation tests, suggesting that the altered transducer activity interferes with CW flagellar rotation. The *cheD* products are deamidated normally, but are excessively methylated, and in fact cause all of the other MCP molecules in the cell to be over-methylated also. It appears that these mutants may have signaling defects which "lock" the transducer into a CCW signaling mode. The excessive methylation probably reflects a futile attempt to cancel this signal. The ability of *cheD* mutations to affect the methylation states of all MCP's could be due either to direct interactions of the transducer molecules or to indirect effects on CheR and CheB activity through feedback control from the flagella (*12*). The *cheD* mutations tend to cluster at two positions in the *tsr* gene, corresponding to the membrane-spanning region that separates the receptor and methylation domains, and to the carboxy-terminal methylation sites. We suggest that the former mutations represent receptor alterations that place the transducer in a permanently excited state regardless of stimulus conditions; the latter mutations may affect the signaling domain directly.

The *tar* gene does not yield mutants analogous to the *cheD* class, but this could simply reflect the fact that the *tsr* gene is expressed at higher levels than the *tar* gene in *E. coli*. Both genes give rise to mutants with CW-biased flagellar rotation, which are probably due to defects in signaling activity. These defects are not as dramatic as those of *cheD* mutants and the CW mutants still exhibit chemotactic responses to stimuli handled by other MCP pathways, although the apparent thresholds are affected by the aberrant rotational pattern of the cell. The CW *tsr* alleles tend to cluster in the same regions of the gene as the *cheD* alleles, suggesting that the CCW and CW rotation patterns are probably due to similar sorts of functional alterations, but with signaling consequences of opposite sign.

X. HYBRID TRANSDUCERS

The *tar* and *tap* loci are located next to one another in the same orien-

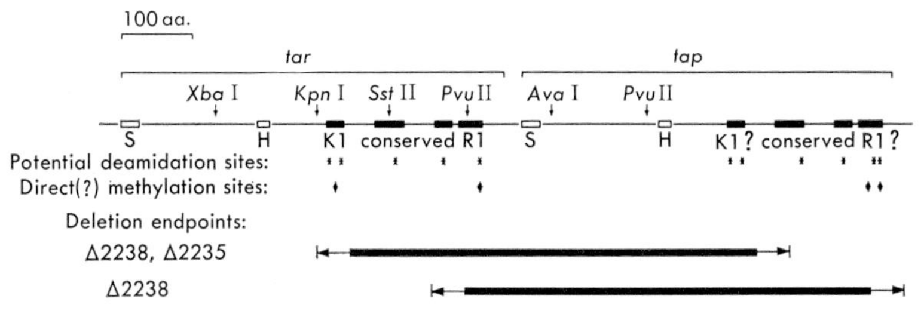

Fig. 2. Structure of *tar-tap* gene fusions. The positions of several deletions that create in-phase fusions of the *tar* and *tap* coding regions are shown relative to landmark features of the transducer molecules: S, hydrophobic signal sequence; H, hydrophobic membrane spanning sequence; K1 and R1, methylated segments (not yet directly demonstrated in Tap); conserved, regions of high homology between all transducers. These positions were established by comparison of fine structure genetic maps with the physical locations of deletion endpoints relative to various restriction endonuclease cleavage sites in the *tar* and *tap* genes (shown at the top of the figure). The limits within which the fusions must lie are indicated by arrows at the bottom of the figure; the solid bars represent the material deleted in the fusion strains.

tation (*15*). We have isolated several deletions that span the *tar-tap* gene border which appear to have arisen by recombination between homologous sequences in the two genes (Fig. 2). These deletions produce in-phase fusions of the *tar* and *tap* coding regions and synthesize hybrid MCP molecules containing the amino end of Tar and the carboxyl end of Tap. One of them (Δ2238) has nearly normal Tar function, consistent with the notion that the amino end of the transducer contains the receptor domain. Evidently in this fusion the carboxy-terminal segment of Tap is able to substitute for the corresponding region of Tar with little effect on transducer function. This fusion appears to contain the carboxyl end from Tap through the R1 segment (see Fig. 2). Two other fusions (Δ2230 and Δ2235) also retain Tar-specific receptor function, but exhibit aberrant signaling properties similar to those seen in *cheD* mutants of *tsr*. These fusions have acquired more of the carboxyl region of Tap, including the R1 methylation sites and at least a portion of the adjacent conserved region (see Fig. 2). Strains with either of these deletions have a severe CCW bias in flagellar rotation and are partially defective in all chemotactic responses. Their fusion proteins are capable of undergoing deamidation, but appear to be excessively methylated. Like *cheD* strains, these defects are dominant to wild type, and suggest that the fusion proteins have altered signaling properties that interfere with sensory responses handled by other MCP pathways. The cause of this phenotype is not yet clear; however, it is interesting to note that the R1 peptide of Tap has several potential methylation sites that are not found in Tar, whereas the K1 peptide of Tar has more sites than its Tap counterpart. The hybrid MCP's could have a greater number of methylation sites than either of the parental molecules, depending on the exact fusion point, and this difference may somehow be responsible for their aberrant signaling properties.

SUMMARY

The methyl-accepting chemotaxis proteins of *E. coli* have many properties characteristic of sensory transducers and appear to be good models for exploring the molecular basis of sensory transduction events. The signaling behavior of these molecules can be modulated in a number of ways, principally by direct interaction with small molecules or liganded binding proteins and by changes in their methylation state. Genetic and biochemical studies of these proteins have led to the following working model of transducer structure and function in bacteria. The transducers are trans-membrane proteins with several different functional activities that appear to correspond to discrete structural domains. Their receptor activity is contained in the amino end, which protrudes into the periplasmic space; their methylation sites are located in the carboxyl end, which remains in the cytoplasm. All transducer species are similar in structure, particularly in their carboxy-terminal half, and may interact directly with one another in the membrane. The signaling activity of the transducers is probably located primarily in the cytoplasmic portion of the molecule, but is clearly influenced by the structure of the amino end as well. Presumably both excitation and adaptation processes modulate transducer signaling properties through reversible conformational changes. The nature of those signals is not yet known.

Acknowledgments

This work was supported by U.S. Public Health Service Grant GM-19559 from the National Institute of General Medical Sciences. M.K.S. was supported by a U.S. Public Health Service predoctoral traineeship from the National Institute of General Medical Sciences.

REFERENCES

1 Berg, H.C. and Brown, D.A. (1972). *Nature* **239**, 500–504.
2 Harayama, S., Palva, E.T., and Hazelbauer, G.L. (1979). *Mol. Gen. Genet.* **171**, 193–203.
3 Hedblom, M.L. and Adler, J. (1980). *J. Bacteriol.* **144**, 1048–1060.
4 Kehry, M.R., Bond, M.W., Hunkapiller, M., and Dahlquist, F.W. (1983). *Proc. Natl. Acad. Sci. U.S.* **80**, 3599–3603.
5 Koiwai, O. and Hayashi, H. (1979). *J. Biochem.* **86**, 27–34.
6 Krikos, A., Mutoh, N., Boyd, A., and Simon, M. (1983). *Cell* **33**, 615–622.
7 Macnab, R.W. and Koshland, D.E., Jr. (1972). *Proc. Natl. Acad. Sci. U.S.* **69**, 2509–2512.
8 Parkinson, J.S. and Hazelbauer, G.L. (1983). In *Gene Function in Prokaryotes*, ed. Beckwith, J., Davies, J., and Gallant, J., pp. 293–318. New York: Cold Spring Harbor Press.
9 Parkinson, J.S. and Revello, P.T. (1978). *Cell* **15**, 1221–1230.
10 Parkinson, J.S., Slocum, M.K., Callahan, A.M., Sherris, D., and Houts, S.E. (1983). In *Mobility and Recognition in Cell Biology*, ed. Sund, H. and Veeger, C., pp. 563–576, Berlin: Walter de Gruyter & Co.
11 Parkinson, J.S. (1980). *J. Bacteriol.* **142**, 953–961.
12 Parkinson, J.S. (1981). In *Genetics as a Tool in Microbiology*, vol. 31, ed. Glover, S.W. and Hopwood, D.A., pp. 265–290, Great Britain: Cambridge University Press.

13 Rollins, C. and Dahlquist, F.W. (1981). *Cell* **25**, 333–340.
14 Sherris, D. and Parkinson, J.S. (1981). *Proc. Natl. Acad. Sci. U.S.* **78**, 6051–6055.
15 Slocum, M.K. and Parkinson, J.S. (1983). *J. Bacteriol.* **155**, 565–577.
16 Springer, W.R. and Koshland, D.E., Jr. (1977). *Proc. Natl. Acad. Sci. U.S.* **74**, 533–537.
17 Springer, M.S., Goy, M.F., and Adler, J. (1979). *Nature* **280**, 279–284.
18 Stock, J.B. and Koshland, D.E., Jr. (1978). *Proc. Natl. Acad. Sci. U.S.* **75**, 3659–3663.
19 Wang, E.A. and Koshland, D.E., Jr. (1980). *Proc. Natl. Acad. Sci. U.S.* **77**, 7157–7161.
20 Yonekawa, H., Hayashi, H., and Parkinson, J.S. (1983). *J. Bacteriol.*, **156**, 1228–1235.

13

BIOCHEMICAL STUDIES ON BACTERIAL CHEMOTAXIS: AN *IN VITRO* SYSTEM OF METHYLATION OF METHYL-ACCEPTING CHEMOTAXIS PROTEINS

SHINSEI MINOSHIMA* AND HIROSHI HAYASHI

Institute of Molecular Biology, School of Science, Nagoya University, Nagoya 464, Japan

Organisms sense their environment and make behavioral responses to environmental change. Such behavior requires the sensing of stimuli, the transduction of generated signals, and the coupling of these signals to a response system. Chemotaxis, the attraction to or repulsion by chemicals involves one of the most primitive sensory transduction systems. In recent years, extensive study of bacterial chemotaxis has proven this phenomenon useful and convenient for the elucidation of molecular mechanisms of sensory transduction. In the process of chemosensory transduction in *Escherichia coli* or *Salmonella typhimurium*, a set of transmembrane proteins (methyl-accepting chemotaxis proteins: MCPs) which can reversibly accept methyl groups on several carboxyl residues play important roles. The MCPs behave as if they were the "interfaces" for transmission of environmental information across the membrane to control cell motility. They gather environmental information in both direct and indirect ways. MCP I and MCP II, for example, are respective chemoreceptors for the attractants serine and aspartate (*8*, *14*),

* Present address: Department of Biology, University of Arizona, Tuscon, Arizona, U.S.A.

while MCP II and MCP III interact with receptor proteins for certain sugars (*10*). Information about receptor occupancy must be processed by the MCPs and integrated and transduced to the motor apparatus to control swimming behavior. With sudden changes in their chemical environment, cells show excitation responses which may last several minutes and then subsequent adaptation to the new environment. The excitation process requires MCP, and the adaptation process requires methylation or demethylation of MCP. Methylation is catalyzed by the *cheR* gene product utilizing *S*-adenosylmethionine as methyl donor. Demethylation is catalyzed by the *cheB* gene product yielding methanol. Thus, MCPs and the enzymes required for MCP modification are indispensable for successful chemosensory transduction. In order to study these reactions under precisely controlled conditions, we constructed an *in vitro* system for the methylation of MCP. Readers wishing a more detailed review of the field are recommended to ref. *11*.

I. CONSTRUCTION OF THE SYSTEM

In vitro methylation of MCP was first demonstrated by Springer and Koshland (*15*) in *S. typhimurium*. We modified their method in applying it to *E. coli* to obtain better results. *E. coli* cells were disrupted using a French press and fractionated by ultracentrifugation. A pellet, consisting mainly of inside-out membrane vesicles, was used as substrate source and is referred to as "membrane". The supernatant, which comes from the cytoplasm, was used as an enzyme source and is referred to as "cytoplasm". The reaction mixture consisted of membrane, cytoplasm and methyl-^{14}C-labeled *S*-adenosylmethionine (AdoMet). Major modifications of the Springer and Koshland procedure were the following three points: first, disruption of cells was carried out by the French press procedure instead of the sonication procedure, which raised the activity of the membrane fraction so that it accepted two to three times more methyl groups. Second, a step for washing the membrane with buffer containing EDTA was inserted, which lowered the background methylation activity associated with the membrane fraction. Third, the assay for the incorporation of radioactive methyl groups onto MCPs was performed after separating the membrane proteins by SDS gel electrophoresis. This procedure eliminated a contamination of radio-

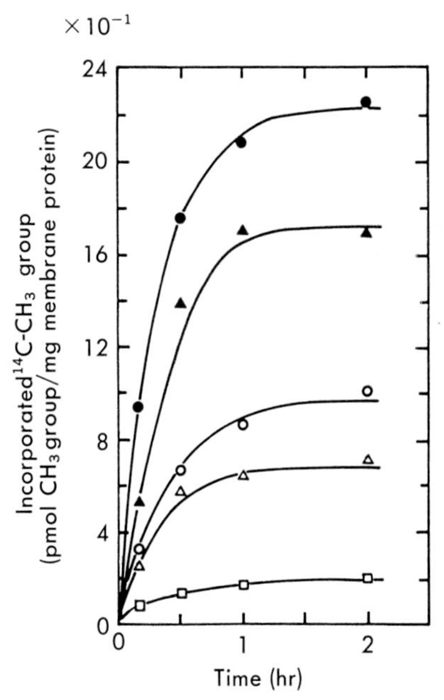

Fig. 1. Time-dependent increase and stationary state of incorporation of methyl groups into MCP. (*12*). Bacteria used were a chemotactically wild type (strain W3110). The reaction mixture contained 7.5 mg of membrane protein, 42 nmol of [^{14}C]AdoMet (48 mCi/mmol), and various amounts of cytoplasmic protein in each ml of buffer. The mixture was incubated at 30°C, and aliquots (45 μl) were withdrawn at the time indicated to determine the amount of radioactivity. The three classes of MCP were assayed without separating them. □, no cytoplasm; △, 1.2 mg/ml cytoplasmic protein; ○, 2.0 mg/ml; ▲, 4.1 mg/ml; ●, 8.3 mg/ml.

activity from the background methylation reaction (*12*). Since the profile on two dimensional gel electrophoresis of proteins methylated in the present system was in good agreement with that in the *in vivo* system (*8*, *9*), the methylating reaction assayed with the *in vitro* system appears to contain no artifacts.

Figure 1 shows typical time courses of the *in vitro* methylation of MCP with varying amounts of cytoplasm in the reaction mixture. The membrane alone showed no significant methylation of MCP. Since no detectable MCP was present in the cytoplasm, we concluded that this system assays the activity contained in cytoplasm which is essential to the methylation of MCP in the membrane.

II. STEADY STATE OF METHYLATION

The MCP methylating reaction eventually reached a steady state where no apparent change in the amount of labeled MCP was found (Fig. 1). It took about 60 min to reach the plateau of the methylation reaction regardless of the amount of cytoplasm added. The steady state level depended on the amount of cytoplasm. The plot of the steady state level against the amount of cytoplasm was almost linear within the range employed in the experiments. The amount of MCP is clearly not a limiting factor of the steady state level, because the amount of membrane added to the reaction mixture was constant, and the level of the plateau was elevated by adding more cytoplasm after the system

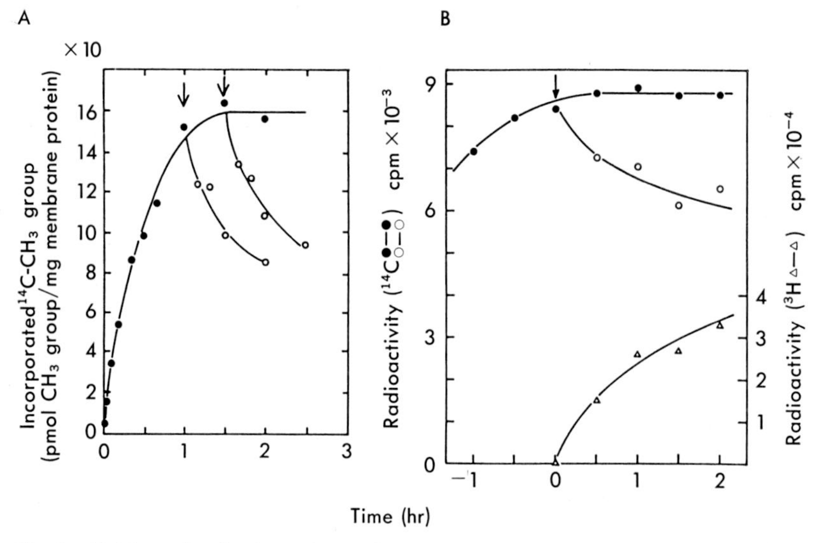

Fig. 2. Evidence for the dynamic steady state of the methylation. A) The reaction mixture contained 3.8 mg/ml of cytoplasmic protein. Aliquots were withdrawn at various times and MCP-methylation was measured. At the times indicated by the arrows, non-radioactive AdoMet was added to a final concentration of 3.8 mM (○). To a control sample (●), the same volume of buffer was added. Change in the volume of reaction mixture (less than 3%) caused a negligible change in its protein concentration. B) The reaction mixture was the same as (A) but contained 9.8 mg/ml cytoplasmic protein. The mixture was divided into two portions at the arrow (time 0), [^{14}C]AdoMet was added to one portion (●), and [^{3}H]AdoMet was added to the other portion (○, △). Incorporation of ^{3}H (△) and ^{14}C (●, ○) into MCP was measured.

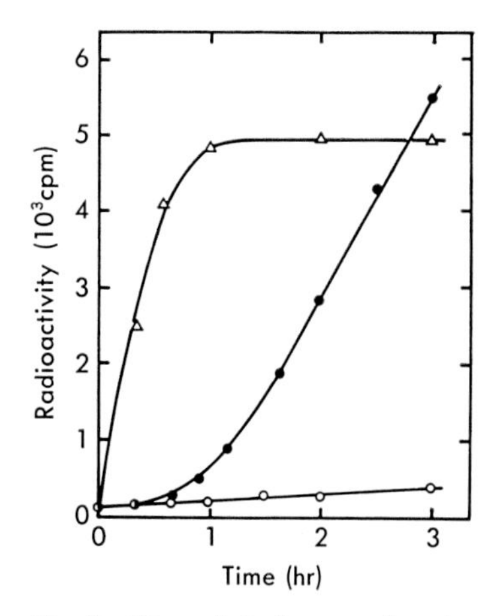

Fig. 3. Demethylation reaction determined *in vitro*. *In vitro* demethylation was assayed by measuring the methanol formed. 75 μl of the reaction mixture was placed in the outer well of a Conway microdiffusion cell and 1 ml of water was placed in the center well. The apparatus was incubated at 25°C for the time indicated after which 1 ml of 10% trichloroacetic acid was added to the outer well to quench the reaction. The apparatus was left for 3 hr at room temperature to establish a liquid-gas equilibrium and the radioactivity of the solution in the center well was determined. ●, membrane+cytoplasm; ○, membrane+buffer (control); △, radioactivity of MCP in the sample.

reached a steady state. It is also evident that the amount of AdoMet did not limit the steady state level. Cytoplasmic activity was not inactivated during this long term incubation because, when the membrane was preincubated with cytoplasm for 60 min and then labeled AdoMet was added to the reaction mixture, the labeling of MCP proceeded at almost the same rate as in the control. We demonstrated that the system was under dynamic equilibrium which has been suggested for the *in vivo* reactions (*9*). After the methylation reaction reached the steady state, a large excess of non-radioactive AdoMet was added to the reaction mixture. Immediately a decrease in radioactivity of MCP was observed indicating that labeled methyl groups had been replaced by cold methyl groups (Fig. 2A). This was further confirmed by a double labeling experiment (Fig. 2B). From these observations, we conclude that the MCP methylating system is in a state of dynamic equilibrium in which methylation and demethylation take place at the same rate and the amount of cytoplasm determines the number of MCP sites available for the exchange of methyl group.

Demethylation of MCP produces methanol *in vivo* (*9*). We examined whether radioactive methanol was produced in the *in vitro* system. The reaction mixture was introduced into Conway microdiffusion cells during the labeling reaction in order to measure the amount of methanol formed. Figure 3 shows a time course of the formation of methanol. Radioactive methanol was formed at a constant rate after a time lag of about 30 min. No significant amount was formed during the lag time, which corresponded to the period when the level of MCP methylation was increasing. Methanol formation results from a turnover of MCP methyl groups. The membrane alone showed no formation of methanol, and a system consisting of membrane and cytoplasm from a *cheR* or *cheB* mutant produced no significant amount of the substance. The formation of methanol at a constant rate confirmed the presence of a dynamic steady state. Its constant formation after a time lag has also been demonstrated in *in vivo* experiments after stimulation with the attractant (*5*). One notable difference between the *in vivo* and *in vitro* systems is that in the latter case the time course is about ten times longer. This may be a consequence of the use of inverted membrane vesicle (reciprocal curvature) and relatively diluted cytoplasm in the *in vitro* system.

III. MULTIPLE METHYLATION OF MCP

The appearance of multiple bands of methylated MCP on SDS-polyacrylamide gels occurs after MCP is stimulated by attractant *in vivo* (*1–4*). In our *in vitro* system, neither the rate nor the final level of methylation at the steady state was affected by addition of attractant. This could be explained by the hypothesis that the membrane preparation is topologically inside out and not permeable to attractants. If this hypothesis holds, there would be no multiple bands of methylated MCP labeled in the system. At various stages in the time course of the *in vitro* labeling of MCP, the MCP was separated by high resolution SDS-polyacrylamide gel electrophoresis (Fig. 4). It is clear that the system produced multiple bands of methylated MCP, suggesting the MCP molecules exist in a stimulated state in the *in vitro* system, as discussed later. The banding pattern of methyl-labeled MCP in the system changed with time and became stationary when the system reached the

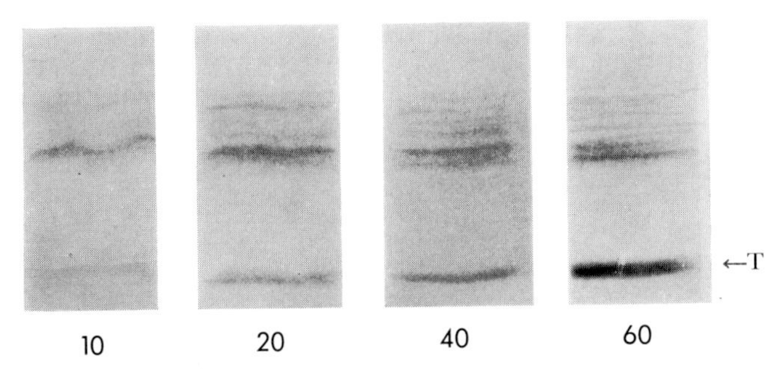

Fig. 4. Time course of labeling of multiple bands of MCP (*7*). MCP was labeled with radioactive methyl groups *in vitro*, separated by high resolution SDS-polyacrylamide gel electrophoresis with reduced bisacrylamide content, and fluorograms obtained. Numbers on the bottom represent the time after addition of [^{14}C]AdoMet in minutes, T indicates the position of peptide elongation factor Tu (*14*).

steady state level. The period of change in banding pattern corresponds to that of increasing methylation level. Radioactivity first appeared in the relatively slow migrating bands while the fastest moving ones were faint or not visible before the steady state was established. Thus, the initial incorporation of radioactivity in the *in vitro* MCP methylation reaction must reflect a net increase in the level of methylation. This conclusion agrees with the results obtained *in vivo*.

IV. EFFECT OF ATTRACTANT IN THE SYSTEM

The present system was not affected by attractant, presumably because the membrane vesicle was inside out. Attractant can be delivered to the membrane vesicle interior by extensive dialysis of the vesicle against a large volume of buffer containing attractant. After prolonged dialysis against various attractant concentrations, the rates of methylation and demethylation were determined. The higher the concentration of attractant in the vesicle, the greater was the rate of methylation and the higher was the level of steady state methylation (Fig. 5). The increase in the level and rate was small but significant and reproducible. The dependence of the rate of methanol formation on attractant concentration at the steady state was not that simple. A semi-logarithmic plot of the rate against the concentration of attractant exhibited a parabola-like shape having a maximum value at about 10^{-5} M (Fig. 5). These

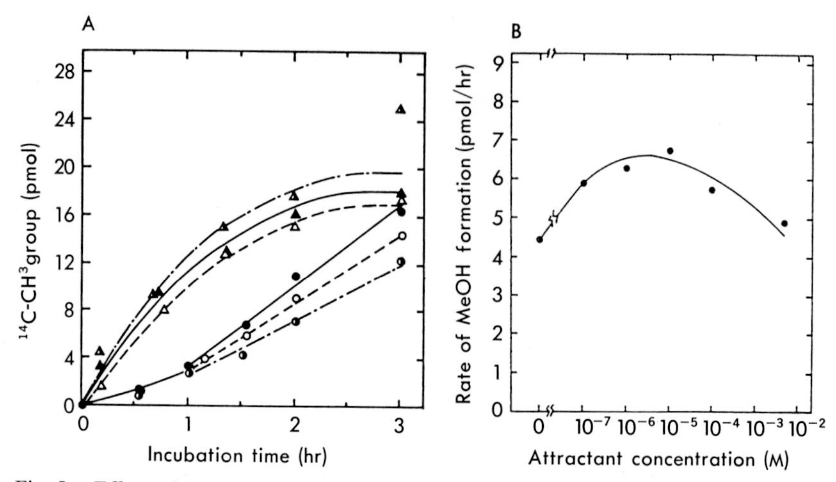

Fig. 5. Effect of attractant on methylation and demethylation *in vitro*. A) The *cheR* membrane was dialyzed against buffer containing various concentrations of L-serine and L-aspartate at 4°C for 60 hr. Methylation and demethylation were assayed by the same method as in Figs. 1 and 3. The cytoplasm was from wild type RP477. Triangles designate the methylation of MCP and circles the methanol formation. ▲, ◑, the membrane was dialyzed against 10^{-3} M serine and aspartate; ▲, ●, 10^{-6} M; △, ○, no attractant. B) A semilogarithmic plot of the rate of methanol formation against the concentration of attractants. Membrane was dialyzed against various concentrations of attractants as in (A).

results showed that our *in vitro* system is sensitive to the concentration of attractant. The demethylation result is very similar to the results of *in vivo* demethylation measurement (*5*).

V. CONCLUSION

The behavior of MCPs in the present *in vitro* methylating system resembles that *in vivo* in the following ways: 1) The level of methylation reaches a steady state, at which it is kept constant by turnover of the methyl groups. 2) Methanol, a hydrolytic product of MCP methyl ester, is formed. 3) The methylation and demethylation reactions require cytosolic factors, namely, the *cheR* and *cheB* gene products, respectively. 4) Methylation and demethylation of MCP is affected by the presence of attractant, presumably by ligand binding to the receptor. 5) Multiply methylated MCP was produced.

The MCP molecules are a marvelous invention of nature which allow *E. coli* to have a short term memory, to detect differential concentration change with time and to adapt to environmental change. The present system provides a powerful tool for the study of receptor methylation and demethylation reactions and the mechanism of their regulation during chemotaxis.

SUMMARY

An *in vitro* system for the assay of methylation and demethylation of MCP was constructed. Consisting of the membrane, the cytoplasm and labeled *S*-adenosylmethionine, the system preserved several characteristics of the *in vivo* system. Multiple methylation and delayed demethylation of MCP were observed and MCP methylation reached a steady-state level determined by the methylation and demethylation reactions in the *in vitro* system. Attractant contained in the reversed membrane vesicles affected the rates of these reactions.

REFERENCES

1 Boyd, A. and Simon, M. (1980). *J. Bacteriol.* **143**, 809–815.
2 Chelsky, D. and Dahlquist, F.W. (1980). *Proc. Natl. Acad. Sci. U.S.* **77**, 2434–2438.
3 DeFranco, A.L. and Koshland, D.E., Jr. (1980). *Proc. Natl. Acad. Sci. U.S.* **77**, 2429–2433.
4 Engstrom, P. and Hazelbauer, G.L. (1980). *Cell* **20**, 165–171.
5 Hayashi, H. and Yonekawa, H. (1982). In *Abstracts of the Twentieth Annual Meeting of the Japanese Society for Biophysics* S213 (in Japanese).
6 Hayashi, H., Koiwai, O., and Kozuka, M. (1979). *J. Biochem.* **85**, 1213–1223.
7 Hayashi, H., Minoshima, S., and Ohba, M. (1982). *J. Biochem.* **92**, 391–397.
8 Hedblom, M.L. and Adler, J. (1980). *J. Bacteriol.* **144**, 1048–1060
9 Kleene, S.J., Hobson, A.C., and Adler, J. (1979). *Proc. Natl. Acad. Sci. U.S.* **76**, 6309–6313.
10 Koiwai, O. and Hayashi, H. (1979). *J. Biochem.* **86**, 27–34.
11 Koshland, D.E., Jr. (1981). *Annu. Rev. Biochem.* **50**, 765–782.
12 Minoshima, S., Ohba, M., and Hayashi, H. (1981). *J. Biochem.* **89**, 411–420.
13 Minoshima, S. and Hayashi, H. (1980). *J. Biochem.* **87**, 1371–1377.
14 Ohba, M., Koiwai, O., Tanada, S., and Hayashi, H. (1979). *J. Biochem.* **86**, 1233–1238.
15 Springer, W.R. and Koshland, D.E., Jr. (1977). *Proc. Natl. Acad. Sci. U.S.* **74**, 533–537.
16 Wang, E.A. and Koshland, D.E., Jr. (1980). *Proc. Natl. Acad. Sci. U.S.* **77**, 7157–7161.

14

RETINAL-CONTAINING PROTEINS IN *HALOBACTERIUM HALOBIUM*

JOHN L. SPUDICH

Department of Anatomy and Department of Physiology and Biophysics, Albert Einstein College of Medicine, New York 10461, U.S.A.

Three retinal-containing proteins have been identified in the membranes of *Halobacterium halobium*: bacteriorhodopsin (*10*), halorhodopsin (*8, 9, 12*), and the "slow-cycling" pigment s-rhodopsin (*1, 17*). Both bacteriorhodopsin and halorhodopsin are light-driven ion pumps, catalyzing transmembrane transport of H^+ and Cl^-, respectively, to hyperpolarize the cell membrane (*10, 12*). s-Rhodopsin does not catalyze electrogenic ion transport and several findings suggest it is a sensory photoreceptor for the cells' phototaxis responses (*1, 17*). It is of considerable interest to compare these three chromoproteins to establish how the same initial event (photoexcitation of retinal) can be coupled to different energy-conserving and signal-generating processes.

Bacteriorhodopsin is one of the best characterized integral membrane proteins in terms of its structure, and much progress has been made on study of its H^+ transport function and photoreaction cycle intermediates (for review, see ref. *19*). In comparison, the structural and functional characterization of halorhodopsin and s-rhodopsin have only just begun. Our understanding of these two pigments derives primarily from study of mutants defective in their functions. In particular, halorhodopsin mutants (*15*) have provided a criterion for the pigment's properties (*1, 7, 14, 15, 17*) and led to the initial detection of s-rhodopsin (*1*). Many published studies of halorhodopsin are confused by the unrecognized presence of s-rhodopsin. This review will summarize the characteristics of halorhodopsin and s-rhodopsin as we now understand them.

I. HALOBACTERIAL RHODOPSIN MUTANTS

The discovery and characterization of retinal pigments other than bacteriorhodopsin in *H. halobium* has depended on ion transport and spectroscopic analysis of several types of mutants, representatives of which are shown in Fig. 1.

Mutants deficient in synthesis of bacteriorhodopsin occur at a high frequency ($>10^{-4}$/generation) (*11*) and are easily recognized by their loss of purple color by screening colonies on agar plates. One example is strain OD2, which derives from the fully pigmented strain S9 (Fig. 1). In strain OD2, as in all other bR^{-} strains so far examined, both halorhodopsin and s-rhodopsin are present (*1*), while bacteriorhodopsin expression is eliminated by a transposable element inserted into the structural gene for bacterioopsin (*2*).

Halorhodopsin-deficient mutants could not be isolated by visual screening but rather required a genetic selection procedure. The key to selecting against halorhodopsin function was to establish conditions

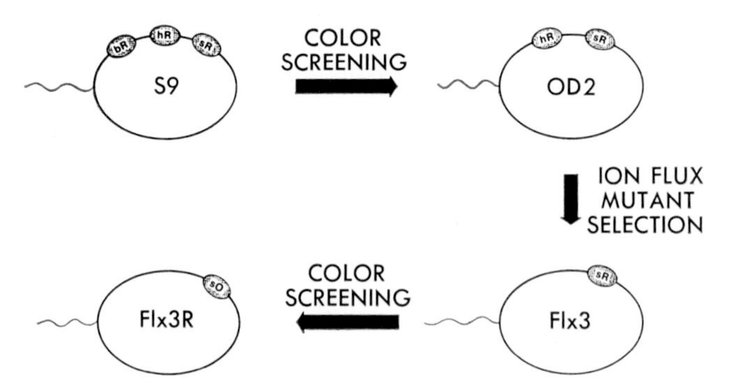

Fig. 1. Representative *H. halobium* mutant strains. bR, bacteriorhodopsin; hR, halorhodopsin, and sR, s-rhodopsin. sO represents s-opsin, the protein moiety of sR. Strains and their isolation are described in ref. *15* and *16*.

in which the ion transport by halorhodopsin was deleterious. This was done by manipulating the cells' transmembrane ion circulation so that photoinduced hyperpolarization by halorhodopsin would overcome the cells' system for regulation of cytoplasmic pH. The mutants generated by the method have altered ion flux properties, many being deficient in halorhodopsin transport. The selection method is described in detail in ref. *16* and in the original report (*15*).

Spectroscopy of wild-type and mutant membranes revealed retinal-dependent photochemical reactions genetically dissociable from halorhodopsin function (*1*). These reactions derive from a separate retinal pigment, s-rhodopsin, the presence of which, as schematized in Fig. 1, is independent of the presence of bacteriorhodopsin or halorhodopsin (*1*). One particular mutant strain, Flx3 (Fig. 1), has been extensively characterized for its content of retinal pigments. Neither bacteriorhodopsin nor halorhodopsin can be detected in Flx3 by ion transport assays (*15, 17*) or by sensitive spectroscopic measurements (*1, 7*) capable of detecting as few as 80 molecules/cell. Flx3 contains s-rhodopsin as the sole detectable retinal pigment in similar amounts as the wild-type strain.

We isolated a retinal-deficient derivative of Flx3 by screening for white (*i.e.*, carotenoid-deficient) colonies, which often show loss of retinal synthesis. The retinal-deficient strain, Flx3R (Fig. 1), lacks photoactive s-rhodopsin but contains the apoprotein (s-opsin), which can generate s-rhodopsin by retinal addition (*17*).

II. SPECTRAL AND CHEMICAL PROPERTIES OF HALORHODOPSIN AND s-RHODOPSIN

The mutants have permitted discrimination of the pigment properties. The main distinguishing characteristics are shown in Table I. It should be noted that several of the properties listed for halorhodopsin in the table differ from those reported prior to the availability of Flx mutants, because the prior studies did not separate halorhodopsin properties from those of s-rhodopsin.

The absorption spectra of the three pigments in their membrane environment has been determined by difference spectroscopy of membrane vesicles (Fig. 2). The absorption spectrum of halorhodopsin was

TABLE I
Properties of *H. halobium* Retinal Pigments

	Bacterio-rhodopsin	Halo-rhodopsin	s-Rhodopsin	References
λ_{max} (4 M NaCl, pH 7)	568 nm	578 nm	587 nm	*17*
Photocycle $t_{1/2}$ (23°C)	7 msec	10 msec	800 msec	*1,17*
Electrogenic?	yes	yes	no	*8,9,10,12,15*
Visible absorption in octylglucoside	Maintained	Maintained	Destroyed	*1,3*
Alkaline shift of λ_{max} (pH 10.8)	to 562 nm	to 410 nm	to 552 nm	*17*
Retinal regulation	Bacterioopsin inducible	Haloopsin inducible	s-Opsin constitutive	*14,20*

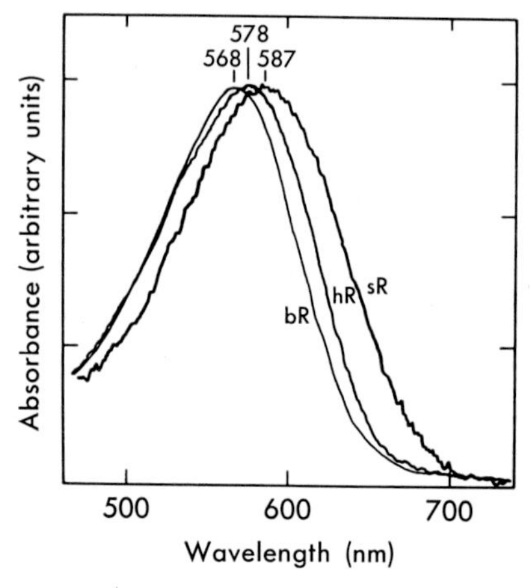

Fig. 2. Absorption spectra of bacteriorhodopsin (bR), halorhodopsin (hR), and s-rhodopsin (sR) in their native states in the *H. halobium* membrane. Redrawn from ref. *17*.

resolved by comparing the absorption of a halorhodopsin-containing strain with a halorhodopsin-deficient mutant derived from it (*15, 17*). Bacteriorhodopsin absorption was similarly resolved by comparing a deficient strain with its pigmented parent (*17*). The s-rhodopsin absorption spectrum was determined as the difference between the absorption of Flx3R membranes with and without retinal (*17*).

The rates of the photochemical reaction cycles of halorhodopsin and s-rhodopsin differ considerably (Table 1). The 10 msec relaxation after flash-induced cycling of halorhodopsin is similar to that of bacteriorhodopsin, whereas the s-rhodopsin 800 msec relaxation is two orders of magnitude slower. The slow step in the s-rhodopsin cycle has interesting consequences, making this molecule photochromic at physiological light intensities. This is explained below with a discussion of its implications to s-rhodopsin's possible function as a phototaxis receptor.

As indicated in Table I, the translocation of Cl^- by halorhodopsin is electrogenic, *i.e.*, its photochemical reactions result in a net transfer of charge across the membrane. This generates a transmembrane electrical potential of the proper polarity (+ outside to − inside) to fuel ion flux-driven processes such as ATP-synthesis, ion antiport, and amino acid transport by the cell membrane. Bacteriorhodopsin also generates a + outside potential (by proton ejection) and both of these pigments therefore contribute to the cells' membrane-linked energy metabolism. The photochemical reactions of s-rhodopsin are non-electrogenic;therefore, mutants such as Flx3 which lack both photoenergizing pigments but contain s-rhodopsin do not hyperpolarize their membranes in response to light (*15*).

The next two properties in Table I, the sensitivities of visible absorption of the pigments to detergent and to alkali, are especially valuable for distinguishing halorhodopsin and s-rhodopsin when they are present in the same membrane. When membranes or vesicles (in 4 M NaCl) are made 1% octylglucoside, halorhodopsin and bacteriorhodopsin are solubilized with little, if any, loss of extinction, while such treatment eliminates s-rhodopsin absorption in the visible (*1*, *3*). The s-rhodopsin visible peak rapidly disappears and a peak at 380 nm is generated upon addition of detergent to a membrane vesicle suspension in 4 M NaCl (R. Bogomolni and J. Spudich, unpublished).

The s-rhodopsin absorption maximum shifts from 587 to 552 nm with a slight (15%) loss of extinction when the membranes are taken from neutral pH to pH 10.8. This alkali-induced transition is completely reversible and provides a method for distinguishing s-rhodopsin from halorhodopsin on absorption criteria, because halorhodopsin when shifted to pH 10.8 loses all extinction above 500 nm, with production of absorbance in the UV/blue region. The alkaline transition of halorhodopsin is also completely reversible. An isosbestic point for the 587 to 552 nm s-rhodopsin transition occurs at 560 nm. Therefore in bacteriorhodopsin-deficient membranes containing the other two pigments the alkali-induced decrease in extinction at 560 nm is a measure of the halorhodopsin content, while the s-rhodopsin content can be calculated from the decrease at 587 nm taking into account the halorhodopsin contribution at that wavelength (*17*). Since these shifts are fully reversible, they provide a non-destructive assay of the amounts of halorhodopsin and s-rhodopsin in membrane preparations containing both pigments. Bacteriorhodopsin undergoes a relatively slight absorption shift to the blue when taken to pH 10.8 (R. Bogomolni, personal communication).

Synthesis of the protein moiety of halorhodopsin is regulated by retinal; *i.e.*, in retinal-deficient strains, the halo-opsin is not expressed and can be induced by adding exogenous retinal to the cells during growth (*14*). On the other hand, the s-opsin is produced constitutively in retinal-deficient mutants and s-rhodopsin can be generated by adding retinal to membranes of a retinal-deficient mutant (*17*). The bacterio-opsin protein apparently is expressed to some extent even in the absence of retinal, but requires retinal for full induction (*20*).

III. ARE THERE MORE THAN 3 RETINAL PIGMENTS IN *H. HALOBIUM?*

Recently Tsuda and coworkers (*23*) resolved slow and fast photoreactions in a bacteriorhodopsin-deficient strain. Although their data alone do not allow one to assign the slow and fast components to separate pigments, the mutant studies (*1*) clearly show that the slow reactions they observe belong to s-rhodopsin and the fast reactions to halorhodopsin. They report that the slow cycling species absorbs to the blue of the fast cycling species (*23*). This does not indicate a pigment different from s-rhodopsin, even though s-rhodopsin absorbs to the red of halorhodopsin (*17*). The apparent conflict derives from their assumption that flash-induced depletion maxima occur at the same wavelength as the maxima of the absorption spectra, which is not true for halorhodopsin (*17*). It is clear the properties of halorhodopsin and s-rhodopsin are sufficient to explain the data reported (*23*).

As will be discussed below, *H. halobium* cells exhibit retinal-dependent phototaxis attraction to red light and repulsion by blue light. These light-dependent motility responses may indicate more retinal pigments are to be discovered. On the other hand, the properties of s-rhodopsin suggest this pigment may mediate both the attractant and repellent responses. This is discussed below after an introduction to the phenomenon of phototaxis of *H. halobium*. If s-rhodopsin mediates the motility responses, then the 3 retinal pigments already identified are sufficient to account for all known retinal-dependent photoprocesses in the cell.

IV. RETINAL PIGMENTS AND PHOTOTAXIS

Motile *H. halobium* cells are attracted to long-wavelength light (>520 nm, for the sake of discussion designated "red") and repelled by shorter-wavelength light (<520 nm, "blue") (*5*). Photostimuli control swimming behavior by modulating the frequency of reversal of swimming direction (*18*). An increase in the intensity of blue light increases the probability of reversal, while decreases in this intensity suppress reversal probability. Red light intensity changes have the inverse effect: increases in intensity suppress, while decreases enhance, reversal probability. This results in a bias of the cell's "random walk" swimming pattern so that its net migration is up gradients of red light and down gradients of blue light. The opposing effects of red and blue light intensity changes on reversal probability add or subtract depending on the directions of the change (*18*).

Studies with the retinal synthesis inhibitor, nicotine, indicated retinal is required for sensory reception of both red and blue stimuli (*4, 13*). Strains genetically defective in retinal synthesis (such as Flx3R in Fig. 1) lack phototaxis and the addition of retinal restores phototaxis responses to both red and blue light (*15, 17*).

Flx3 and similar mutants which lack both bacteriorhodopsin and halorhodopsin exhibit phototaxis responses (*15*) with red and blue sensitivities essentially the same as those of related strains containing bacteriorhodopsin and halorhodopsin (*1*). Therefore, at least one sensory rhodopsin, a retinal pigment distinct from the energy-harvesting rhodopsin-like pigments, must exist in the cells to mediate the phototaxis.

Is s-rhodopsin a sensory receptor for phototaxis? It absorbs in the red and several findings suggest it may mediate the attractant (*i.e.*, red light) sensitivity. First, s-rhodopsin is the only photoactive retinal pigment detectable in fully phototactically sensitive Flx mutants such as

Flx3. Second, retinal-deficient mutants (*e.g.*, Flx3R, Fig. 1) lack phototaxis responses, which are restored by adding retinal. Therefore the phototaxis receptor opsin(s) must be constitutive, like the s-opsin in these cells. Third, retinal addition to Flx3R cells generates s-rhodopsin and phototaxis sensitivity with similar kinetics and at similar concentrations (B. Ehrlich, C. Schen, and J. Spudich, in preparation), and we detect only a single absorbing species (s-rhodopsin) generated by retinal addition (*17*). These correlations do not prove s-rhodopsin involvement in phototaxis, but strongly recommend it for consideration as a sensory receptor.

The 587 nm absorption maximum of the pigment does not fit precisely with the action spectra for phototactic attraction measured by Hildebrand and coworkers, who report a single attractant maximum at 565 nm in wild-type (*5*), as well as in a bacteriorhodopsin-deficient strain, ET15 (*6*). However, in another bacteriorhodopsin-deficient strain, L33, they report two attractant maxima, one at 565 nm and a second at 590 nm (*22*). Halorhodopsin and s-rhodopsin are both present in both ET15 and L33, but neither of these pigments has its absorption maximum at 565 nm (*17*). Takahashi and Kobatake (*21*) used automated computer-linked tracking of cells of strain S9 to compare the sensitivities at 573 nm and 585 nm, and found the latter wavelength to be more active as an attractant. This result is inconsistent with the 565 nm maximum, but consistent with s-rhodopsin mediation of the attraction.

More studies are needed to understand the action spectra. The action spectra for Flx mutants would be especially valuable since these are free of possible complicating effects of photoexcitation of bacteriorhodopsin and halorhodopsin during attractant stimulation. While the 565 nm phototaxis sensitivity peak *vs.* the 587 nm s-rhodopsin absorption is difficult to explain, given the other evidence the mismatch does not seem sufficient to rule out s-rhodopsin as an attractant receptor.

What pigment mediates the repellent (blue) response, which shows an action spectrum maximum near 370 nm (*5*)? It is possible that s-rhodopsin mediates this repellent sensing due to its photochromic property, which produces a photoactive intermediate absorbing at 373 nm.

The photochromicity of s-rhodopsin derives from the slow step in its photocycle. In the dark, the pigment absorbs red light (λ_{max}=587 nm and therefore we called this state sR_{587}). When sR_{587} is photoexcited it converts to a species absorbing maximally at 373 nm, which returns very slowly (the 800 msec step) to the red-absorbing form. Because of the fast rise and slow decay of S_{373}, this intermediate accumulates in significant amounts under physiological illumination conditions, producing a photostationary state mixture of the two species sR_{587} and S_{373}. Flashing this photostationary state mixture with near UV light reveals that photoexcitation of S_{373} rapidly converts it to sR_{587} (*1*). Because both S_{373} and sR_{587} can be interconverted by light, the composition of the photostationary state is sensitive to changes in both light intensity and spectral composition. The pigment therefore behaves as a photochromic system, sensitive to both light intensity and color. s-Rhodopsin shares this property with color-discriminating photosensors such as phytochrome and also with invertebrate rhodopsins.

If both the attractant and repellent photoreception derive from the same pigment, the red/blue opposition in the taxis behavior would be readily understandable in terms of the opposing effects of red and blue light on the photostationary state. The simplicity of the photochromicity hypothesis for taxis makes it very attractive, and the presence of two opposing photoactive forms of s-rhodopsin is consistent with the pigment functioning in this way.

SUMMARY

Three photochemically reactive retinal-containing pigments have been identified in the membranes of *H. halobium*. Two of these, bacteriorhodopsin and halorhodopsin, are light-driven electrogenic ion pumps which energize the cell by hyperpolarization of the cytoplasmic membrane. The third retinal pigment is active in the same spectral region as bacteriorhodopsin and halorhodopsin, but has a slow (sec) photoreaction cycle compared to the fast (msec) cycles of the photoenergizing pigments. To indicate its slow cycling, the new pigment has been designated s-rhodopsin. Unlike bacteriorhodopsin and halorhodopsin, s-rhodopsin is not an electrogenic ion pump. It may function as a sensory receptor for phototaxis since mutants containing only s-rhodopsin still show the phototaxis responses of the wild-type and the properties of s-rhodopsin suggest it could mediate these responses. All three pigments have very similar absorption spectra. Mutants lacking one or more of the pigment activities have provided a way to discriminate their individual spectral and chemical properties, which are compared in this review.

Acknowledgments
This work was supported by NIH grant GM27750, Jane Coffin Childs Memorial Fund Project Grant No. 360, and an Irma T. Hirschl Trust Career Scientist Award to the author.

REFERENCES

1 Bogomolni, R.A. and Spudich, J.L. (1982). *Proc. Natl. Acad. Sci. U.S.* **79**, 6250–6254.
2 DasSarma, S., RajBhandary, U.L., and Khorana, H.G. (1983). *Proc. Natl. Acad. Sci. U.S.* **80**, 2201–2205.
3 Dencher, N.A. and Heyn, M.P. (1978). *FEBS Lett.* **96**, 322–326.
4 Dencher, N.A. and Hildebrand, E. (1979). *Z. Naturforsch.* **34**, 841–847.
5 Hildebrand, E. and Dencher, N. (1975). *Nature* **257**, 46–48.
6 Hildebrand, E. and Schimz, A. (1983). *Photochem. Photobiol.* **37**, 581–584.
7 Krupinski, J., Spudich, J.L., and Hammes, G.G. (1983). *J. Biol. Chem.* **258**, 7964–7967.
8 Macdonald, R.E., Greene, R.V., Clark, R.D., and Lindley, E.V. (1979). *J. Biol. Chem.* **254**, 10986–10994.
9 Matsuno-Yagi, A. and Mukohata, Y. (1980). *Arch. Biochem. Biophys.* **199**, 297–303.
10 Oesterhelt, D. and Stoeckenius, W. (1973). *Proc. Natl. Acad. Sci. U.S.* **70**, 2853–2857.
11 Pfeifer, F., Weidinger, G., and Goebel, W. (1981). *J. Bacteriol.* **145**, 375–381.
12 Schobert, B. and Lanyi, J.K. (1982). *J. Biol. Chem.* **257**, 10306–10313.
13 Sperling, W. and Schimz, A. (1980). *Biophys. Struct. Mech.* **6**, 165–169.
14 Spudich, E.N., Bogomolni, R.A., and Spudich, J.L. (1983). *Biochem. Biophys. Res. Commun.* **112**, 332–338.
15 Spudich, E.N. and Spudich, J.L. (1982). *Proc. Natl. Acad. Sci. U.S.* **79**, 4308–4312.
16 Spudich, J.L. (1983). In *Information and Energy Transduction in Biological Membranes*, ed. Helmreich, E., Bolis, L., and Passow, H., New York: Alan R. Liss, Inc., in press.
17 Spudich, J.L. and Bogomolni, R.A. (1983). *Biophys. J.* **43**, 243–246.
18 Spudich, J.L. and Stoeckenius, W. (1979). *J. Photobiochem. Photobiophys.* **1**, 43–53.
19 Stoeckenius, W. and Bogomolni, R.A. (1982). *Annu. Rev. Biochem.* **52**, 587–616.
20 Sumper, M. and Herrmann, G. (1976). *FEBS Lett.* **69**, 149–152.
21 Takahashi, T. and Kobatake, Y. (1982). *Cell Struct. Funct.* **7**, 183–192.
22 Traulich, B., Hildebrand, E., Schimz, A., Wagner, G., and Lanyi, J.K. (1983). *Photochem. Photobiol.* **37**, 577–579.
23 Tsuda, M., Hazemoto, N., Kondo, M., Kamo, N., Kobatake, Y., and Terayama, Y. (1980). *Biochem. Biophys. Res. Commun.* **108**, 970–976.

15

THERMOSENSORY TRANSDUCTION IN *ESCHERICHIA COLI*

TAKAFUMI MIZUNO, KAYO MAEDA*, AND YASUO IMAE

Institute of Molecular Biology, Faculty of Science, Nagoya University, Nagoya 464, Japan

Because temperature is an important factor in cellular activity, most organisms have the ability to sense and respond to a thermal stimulus. However, thermosensory cells even in higher organisms generally do not form highly specialized sensory organs such as noses for chemosensory and eyes for photosensory, so that very little is known of the thermosensing mechanism. We have been studying the thermosensory transducing system of *Escherichia coli*, since we can easily apply genetic and biochemical techniques to its analysis. The chemosensory transducing system in *E. coli* has also been extensively studied so that we can use the accumulated knowledge and mutants of the chemosensory system for the analysis of thermosensory system in *E. coli*. Thus, the thermosensory transducing system in *E. coli* provides an excellent model of the molecular mechanism of thermosensing in higher organisms as well as in bacteria.

* Present address: Max-Planck-Institute for Medical Research, Heidelberg, West Germany.

I. BASIC PROPERTIES OF THE THERMOSENSORY SYSTEM IN *E. COLI*

Similar to their response to chemical stimuli, *E. coli* cells respond to thermal stimuli by changing their swimming patterns (*1*, *7*). At constant temperatures, cells show random swimming which is a combination of smooth swimming and intermittent tumbling. A temperature increase induces smooth swimming by fixing the direction of flagellar rotation counter-clockwise, and a temperature decrease induces tumbling by fixing the direction of flagellar rotation clockwise. After a while, the cells begin to adapt to the new temperature and their pattern gradually returns to random swimming. Thus, for *E. coli*, a temperature increase corresponds to an increase in the attractant concentrations and a temperature decrease to an increase in the repellent concentrations.

When the size of temperature change (dT) was fixed but the rate of temperature change (dT/dt) was varied, the size of the thermoresponse was linearly increased with a logarithmic increase in the rate of temperature change (Fig. 1). This result clearly indicates that *E. coli* cells sense not the dT but the dT/dt as the thermal stimulus. The threshold

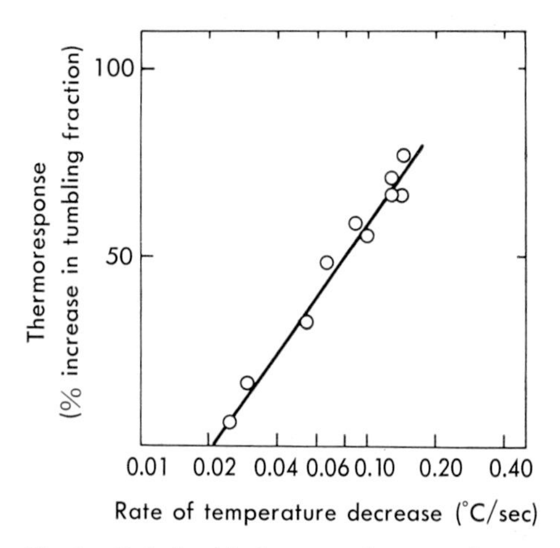

Fig. 1. Relationship between the rate of temperature change and the size of thermoresponse in *E. coli*. Wild-type cells (W3110) were exposed to a linear decrease in temperature from 35°C with various changing rates. Percent increase in the tumbling fraction in total swimming cells was measured 15 sec after the temperature change.

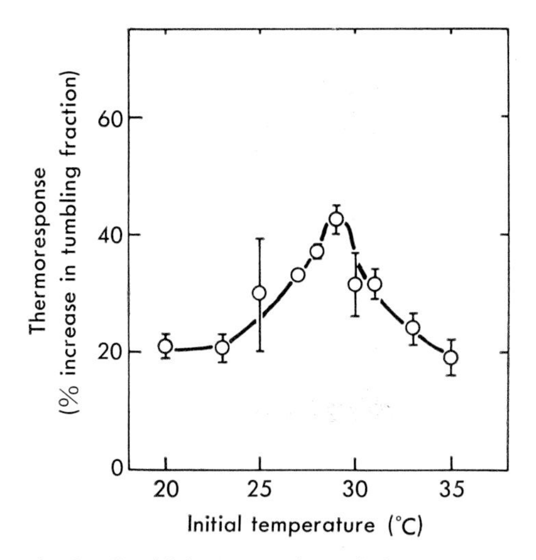

Fig. 2. Sensitivity curve of *E. coli* thermosensory system. Wild-type cells (W3110) were exposed to a linear decrease in temperature at a constant rate of 0.05°C/sec, and the starting point of the temperature change was varied as indicated. Percent increase in the tumbling fraction was measured 15 sec after the temperature change.

thermal stimulus is estimated to be about 0.02°C/sec. When the dT/dt was fixed at 0.05°C/sec but the starting temperature for the temperature change was varied, the size of the thermoresponse was maximum at the starting temperature of about 29°C (Fig. 2). Thus, the sensitivity of the *E. coli* thermoreceptor is optimal at around 29°C.

II. THERMOSENSORY TRANSDUCING PATHWAY

In the case of chemosensory transduction in *E. coli*, there are three major pathways in which the methyl-accepting chemotaxis proteins (MCPs) have a key role. The MCPI-pathway is for the sensing of L-serine and L-alanine, the MCPII-pathway is for the sensing of L-aspartate and maltose, and the MCPIII-pathway is for the sensing of D-ribose and D-galactose.

At least a part of the thermosensory transducing pathway was suggested to overlap with the chemosensory transducing pathway, since general-non-chemotactic mutants were completely lacking in thermoresponse ability (*6*, *7*). We therefore investigated the relationship be-

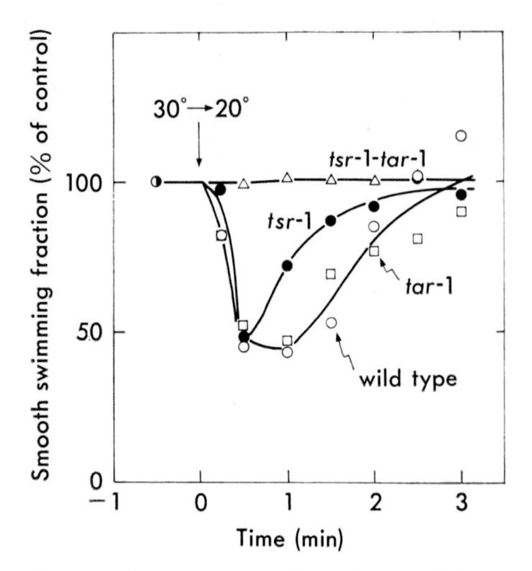

Fig. 3. Thermoresponse in various MCP mutants. Temperature of the cell suspension was decreased from 30°C to 20°C at the straight arrow. Changes in the smooth swimming fraction were measured. ○, wild-type, W3110; ●, MCPI-deficient mutant, AW518 (*tsr*-1); □, MCPII-deficient mutant, AW539 (*tar*-1); △, MCPI-MCPII-deficient mutant, AW569 (*tsr*-1-*tar*-1).

tween various MCP-pathways and the thermosensory transducing pathway by measuring the size of thermoresponse in various MCP mutants. As shown in Fig. 3, mutants having double defects in MCPI and MCPII showed no thermoresponse, indicating that the thermal stimulus is processed through the MCPI- or the MCPII-pathway or both. Mutants defective in MCPI showed clearly weaker thermoresponse than the wild-type strains but other mutants defective in MCPII or MCPIII showed normal thermoresponse. These results suggest that the thermal stimulus is mainly processed through the MCPI-pathway. Consistent with this result, the thermoresponse was strongly inhibited by the adaptation of the cells to an MCPI-specific attractant, L-serine. MCPII-specific attractants such as L-aspartate or MCPIII-specific attractants such as D-galactose caused only a slight inhibition of the thermoresponse. Thus, it is concluded that the thermal stimulus is mainly processed through the MCPI-pathway.

Recent evidence indicates that the MCPs have a dual function as chemoreceptor and chemosensory transducer (*2*, *4*, *8*, *10*). It is therefore very likely that MCPI itself is the main thermoreceptor in *E. coli* and that the sensory signals are produced as a result of the conformational change of MCPI upon temperature changes.

III. CONDITIONAL INVERSION OF THE THERMORESPONSE

A deletion mutant of MCPI (*Δtsr*) was found to show weaker thermoresponse than the wild-type strains but to still have some ability to respond to the thermal stimulus. This result indicates that, in addition to the MCPI-pathway, the MCPII-pathway is somewhat concerned with the thermosensory transduction in *E. coli*, since the only pathways for the thermosensory transduction are the MCPI- and MCPII-pathways as described earlier. During the investigation of this residual thermoresponse in MCPI-deficient mutant we found a strange but interesting phenomenon: under a certain condition the residual thermoresponse could be inverted.

Mutants completely lacking in MCPI, *tsr*-1 or *Δtsr* showed a weak thermoresponse but its direction was the same as that of wild-type

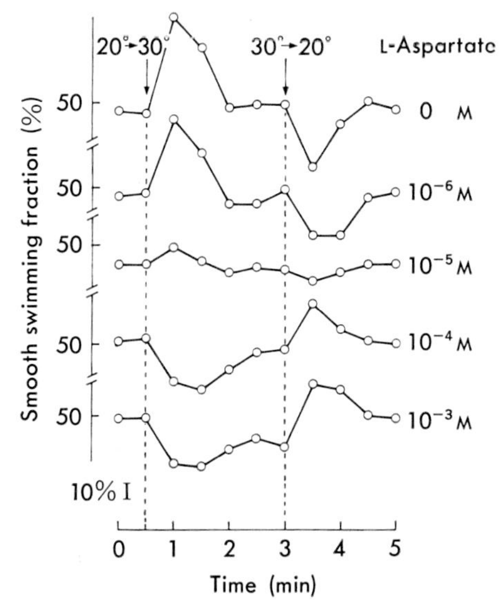

Fig. 4. Relationship between inversion of the thermoresponse and concentration of L-aspartate. An MCPI-deficient mutant (AW518) was incubated at 35°C for 40 min with various concentrations of L-aspartate. Temperature was then changed as indicated by the arrows and the increase in the smooth swimming fraction was measured.

strains. However, when the mutant cells were adapted to MCPII-specific attractants such as L-aspartate and maltose, the thermoresponse direction was completely inverted; a temperature increase induced temporary tumbling and a temperature decrease induced temporary smooth swimming.

When the concentration of L-aspartate was varied, the thermoresponse was inverted in the presence of 10^{-4} M or higher concentrations of L-aspartate (Fig. 4). Since L-aspartate specifically and directly interacts with MCPII and the dissociation constant is estimated to be about 5×10^{-6} M (*3*), MCPII saturation by L-aspartate seems essential to induce the inverted thermoresponse. Other MCPII-specific attractants were also effective in inducing the inverted thermoresponse, whereas the attractants not specific to MCPII or the repellents specific to the MCPII-pathway were ineffective. The inverted thermoresponse was also observed in wild-type strains only when the cells were simultaneously adapted to L-serine and L-aspartate. Thus, the thermoresponse is inverted when the MCPI-pathway is blocked by either mutations or adaptation to L-serine and the MCPII-pathway is blocked by adaptation to MCPII-specific attractants.

In contrast, mutants defective in MCPII showed no thermoresponse inversion, even after adaptation of the cells to L-serine; the thermoresponse was simply inhibited by L-serine as in the case of wild-type cells. These results indicate that MCPII has a key role in thermoresponse inversion and it is concluded that, in addition to MCPI, MCPII also has a minor but definite role in *E. coli* thermosensing.

IV. METHYLATION LEVEL OF MCPS AND THERMOSENSING

Adaptation to chemical stimuli is caused by changes in the methylation level of relevant MCPs (*9*). The results presented in the previous section show that both the MCPI- and MCPII-pathways are involved in thermosensory transduction. Consistent with this, the methylation levels of both MCPI and MCPII showed clear changes upon temperature changes as shown in Table I. In a condition where the cells showed a normal thermoresponse, a temperature increase caused an increase in the methylation level of both the MCPs and a temperature decrease caused a decrease in this level. With MCPI-deficient mutants, an in-

TABLE I
Changes in the Methylation Level of MCPs with a Temperature Decrease from 30°C to 20°C.

Strain	Attractant added	Change in the methylation level of MCPI	Change in the methylation level of MCPII
		%	%
RP5698 (*Δtsr*)	L-aspartate, 10^{-6} M	—	−24
	L-aspartate, 10^{-3} M	—	+30
RP4324 (*tar*52 *Δ*1)	L-serine, 10^{-6} M	−25	—
	L-serine, 10^{-3} M	0	—

Cells were incubated at 30°C for 30 min with [*methyl*-^{3}H]-L-methionine in the presence of the attractant indicated. Then, half of the cell suspension was transferred to 20°C and the remaining half was incubated at 30°C and used as a control culture. Methylation levels of MCPI or MCPII in the two samples were measured 20 min later (*9*). The difference, namely, the change in methylation level of MCPI or MCPII with the temperature decrease, is expressed as a percentage of the methylation level of the MCP in the control culture.

verted change in the methylation level of MCPII was observed in the condition where the thermoresponse was inverted. In the case of MCPII-deficient mutants, however, no change in the methylation level of MCPI was observed in the condition where the thermoresponse was completely inhibited. Thus, the direction of the thermoresponse is always parallel with the direction of change in the methylation level of MCPI and MCPII, regardless of the direction of temperature change. The changes in methylation level of MCPs are shown to be associated with the adaptation of the cells to thermal stimulus and, therefore, it is concluded that *E. coli* cells sense the thermal stimulus at or before the step of MCPs.

V. SIMPLE MODEL FOR THERMORECEPTION

Based on the results presented above, we can propose a model to simply explain the normal and inverted thermoresponses in *E. coli* (Fig. 5). Assumptions are (i) that the thermoreception takes place as a result of conformational changes in both MCPI and MCPII and (ii) that the conformational changes are affected by the methylation level of these MCPs. With a temperature change, both the MCPI and MCPII which have low methylation levels produce positive signals, whereas the highly methylated MCPI produces no signal and the highly methylated MCPII produces negative signals. With this model, it is easily explained that

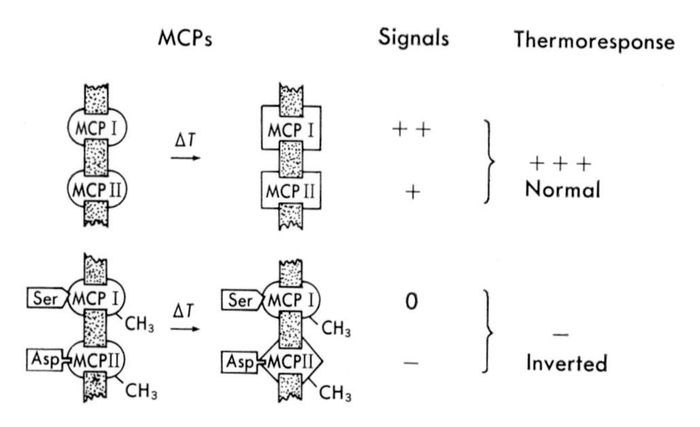

Fig. 5. A simple model explaining the mechanism of thermoreception and the inverted thermoresponse in *E. coli*. MCPs are transmembrane proteins and are assumed to be thermoreceptor proteins. With changes in temperature they show conformational changes and produce sensory signals, although the sign of the signals differs depending on the methylation level of MCPs. See text in detail. +, positive signal; –, negative signal; 0, no signal.

wild-type cells in the absence of any attractant show a normal thermoresponse, whereas cells adapted simultaneously to L-serine and L-aspartate show an inverted thermoresponse. Furthermore, it is also shown that MCPI-deficient mutants show an inverted thermoresponse only in the presence of L-aspartate and that MCPII-deficient mutants never show it.

SUMMARY

E. coli cells show a transient response to thermal stimulus by changing their swimming patterns. The cells sense the changing rate of temperature (dT/dt) as the thermal stimulus and the threshold is about 0.02°C/sec. Sensitivity of the cells to the thermal stimulus is optimal at around 29°C.

Using various MCP mutants, the thermal stimulus was found to be processed through both the MCPI- and MCPII-pathways, although the former was the more dominant. When the methylation level of MCPI was increased, the size of the thermoresponse was clearly decreased and, in the absence of MCPI, the methylation level increase of MCPII resulted in the inversion of the thermoresponse. These results indicate that the signaling properties of these MCPs to thermal stimulus are modulated differently by their methylation. It is very likely, therefore, that both MCPI and MCPII themselves are the thermoreceptor in *E. coli* and that thermoreception is accomplished through their conformational changes upon temperature changes.

Acknowledgments
We would like to thank Prof. J. Adler of the University of Wisconsin and Prof. J. S. Parkinson of the University of Utah for providing various *E. coli* mutants. We would also like to thank Prof. F. Oosawa of our Institute for helpful discussions and encouragement.

This work was supported in part by a Grant-in-Aid from the Ministry of Education, Science, and Culture of Japan.

REFERENCES

1 Adler, J. (1976). *Sci. Am.* **234**, No. 4, 40–47.
2 Boyd, A., Kendall, K., and Simon, M.I. (1983). *Nature* **301**, 623–626.
3 Clarke, S. and Koshland, D.E., Jr. (1979). *J. Biol. Chem.* **254**, 9695–9702.
4 Hedblom, M.L. and Adler, J. (1980). *J. Bacteriol.* **144**, 1048–1060.
5 Krikos, A., Mutoh, N., Boyd, A., and Simon, M.I. (1983). *Cell* **33**, 615–622.
6 Maeda, K. and Imae, Y. (1979). *Proc. Natl. Acad. Sci. U.S.* **76**, 91–95.
7 Maeda, K., Imae, Y., Shioi, J.-I., and Oosawa, F. (1976). *J. Bacteriol.* **127**, 1039–1046.
8 Russo, A.F. and Koshland, D.E., Jr. (1983). *Science* **220**, 1016–1020.
9 Springer, M.S., Goy, M.F., and Adler, J. (1979). *Nature* **280**, 279–284.
10 Wang, E.A. and Koshland, D.E., Jr. (1980). *Proc. Natl. Acad. Sci. U.S.* **77**, 7157–7161.

16

RECONSTITUTION OF IONIC CHANNELS INTO LIPID BILAYER MEMBRANES

RAMON LATORRE*[1] AND DALE BENOS*[2]

*[1]*Departamento de Biología, Facultad Ciencias Básicas y Farmaceúticas, Universidad de Chile, Santiago, Chile and* *[2]*Department of Physiology and Biophysics Laboratory of Human Reproduction and Reproductive Biology, Harvard Medical School, Massachusetts 02115, U.S.A.*

Aqueous pores in membranes have been discussed in the physiological literature since Bernstein (*5*) first proposed his membrane theory (see (*2*)). Pores provide a convenient way to transfer ions and polar solutes through biological membranes. However, they have proved to be much more important than a simple mechanism to communicate the intracellular and extracellular media. Many physiological functions such as the conduction of nerve impulses, the regulation of pacemaker activity, the heartbeat, hormone secretion and fertilization require the proper and exquisitely timed changes in ionic channel mediated conductances (*14*, *19*, *33*, *34*).

The most compelling evidence for the existence of ionic channels in cell membranes comes from the direct recording of quantal current jumps generated by the spontaneous opening and closing of individual transport proteins (*39*). Single channel activity has been measured in two ways: either by the patch clamp technique (*16*) or in planar lipid bilayer membranes doped with various channel forming molecules (*23*, *24*, *29*). These techniques allow direct study of single channel properties in an environment of essentially controllable composition and electrochemical potential.

Isolation of pure ion conducting proteins from natural membranes and subsequent incorporation into planar lipid bilayers is a field that has flourished in recent years. Functional reconstitution of channel forming proteins originating from sarcoplasmic reticulum (*28*), transverse tubules (*25*), axon membranes (*13*), *Escherichia coli* (*4*, *49*), mitochondria (*9*, *46*), brain (*21*) and now even epithelial membranes (*43*, *44*) has been successful.

In this paper we will review the three main techniques that have been used either to reconstitute or to transfer ionic channels from native to artificial membranes. First, we will discuss the incorporation of channels by means of fusion of membrane vesicles with planar lipid bilayers. Second, we consider the formation of planar bilayers from monolayers containing the channel-forming proteins. Mixtures of the fusion and folded bilayer methods are included in both of these two techniques for the sake of simplicity. Third, we examine a recently developed method that consists of constructing bilayers from monolayers on patch-clamp pipettes. Our objective is to discuss the advantages and disadvantages of each method using specific examples of channels studied in this manner.

I. THE FUSION METHOD

1. *Conditions for Fusion: Channels Can Be Safely Transferred from Native Membranes to Planar Bilayers*

Table I presents a compilation of the different ionic channels that have been incorporated into planar bilayers by the fusion method first implemented by Miller and Racker (*31*) and Miller *et al.* (*30*). These investigators reported that three experimental conditions must be met for appreciable fusion of membrane vesicles to solvent-containing planar bilayers to occur: a) an osmotic gradient across the planar bilayer, the vesicle-containing (*cis*) side being hyperosmotic; b) millimolar concentrations of Ca^{2+} in the *cis* electrolyte solution; and c) the presence of negatively-charged phospholipids in the planar bilayer. We believe, however, that the only absolute requirement for fusion is the osmotic gradient (see below). Nonetheless, under these experimental conditions,

TABLE I
Ionic Channels Incorporated into Lipid Bilayers

Technique	Membrane origin	Channel type and conductance	Ionic conditions	Reference
Fusion	Rat Brain	Na^+ (30 pS)	0.5 M NaCl	*21*
	Cultured A6 kidney cells	Na^+ (4–60 pS)	0.2 M NaCl	*43, 44*
	Transverse tubule	K^+ (230 pS)[A]	0.1 M KCl	*25*
	Cardiac sarcolemma	K^+ (95 pS)	0.1 M KCl	*10*
		K^+ (28 pS)	0.1 M KCl	*10*
	Cultured A6 kidney cells	K^+ (30 pS)[B]	0.2 M KCl	
	Sarcoplasmic reticulum	K^+ (140 pS)	0.1 M KCl	*32*
	Smooth muscle	K^+ (292 pS)[C]	0.1 M KCl	
	Cardiac sarcolemma	Cl^- (55 pS)	0.1 M KCl	*10*
	Torpedo electroplax	Cl^- (10 and 20 pS)[D]	0.2 M NaCl	*18, 53*
	Torpedo electroplax	ACh channel, cationic (95 pS)	1.0 M NaCl	*6*
Folded bilayers	Mitochondria	Unselective (450 pS)	0.1 M KCl	*9, 46*
	Lobster axon	K^+ (85 pS)	0.5 M KCl/0.1 M KCl[E]	*13*
	E. coli	Unselective (140–480 pS)	0.1 M NaCl	*49*
	Torpedo electroplax	Cationic (28 pS)	0.3 M NaCl	*36*
"Dip" Bilayers	*Torpedo* electroplax	ACh channel, cationic (40 pS)	0.45 M NaCl	*36, 51*
	Cardiac sarcolemma	Cl^- (70 pS)	0.1 M KCl/0.05 M KCl[E]	*11*
		K^+ (35 pS)	0.1 M KCl/0.05 M KCl[E]	*11*
	Lobster axon	K^+ (85 pS)	0.1 M KCl/0.05 M KCl[E]	*11*
		K^+ (26 pS)	0.1 M KCl/0.05 M KCl[E]	*11*
	Paramecium cilia fragments	Cationic (10–14 pS)	0.1 M KCl	*54*

A. Ca^{2+}-activated K^+ channel. B. Olans, L. and Benos, D., unpublished observations. C. Cecchi, X., Alvarez, O., Wolff, D., and Latorre, R., unpublished observations. D. This Cl^- channel has three conductance states: 0 pS (closed), 10 and 20 pS. E. Concentrated side is that to which vesicles were added.

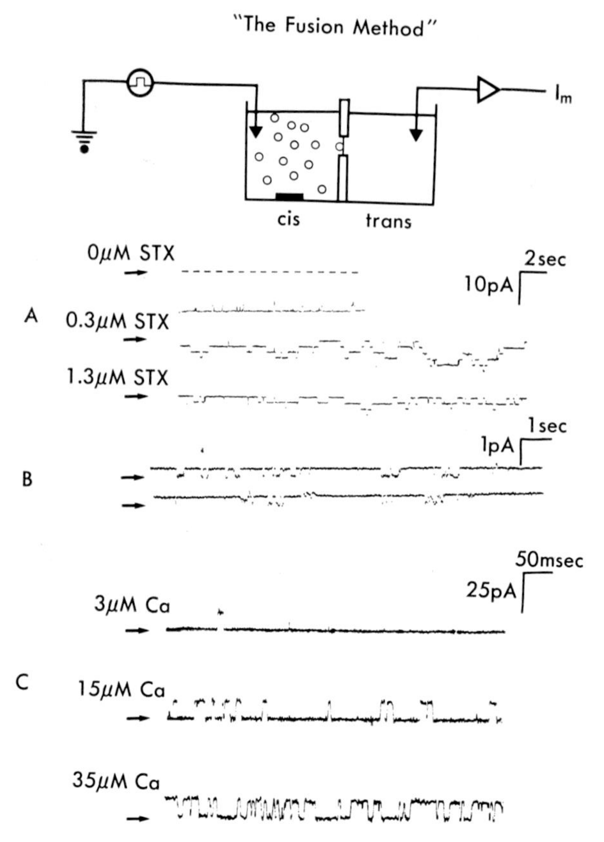

Fig. 1. Schematic representation of the fusion method. In this method bilayers can be formed either with the Mueller and Rudin technique *(38)* or from lipid monolayers spread at the air/water interface *(37)*. Vesicles are added to one side of the bilayer only and fusion is usually induced by establishing an osmotic gradient across the bilayer in conjunction with constant stirring. Some typical single channel records observed when using this technique are also shown. (A) Sodium channels from crude rat brain axolemma. Channels were incorporated in the presence of *trans* batrachotoxin, followed by addition of saxitoxin (STX) to the *cis* compartment. In this experiment, the membrane potential was held at −60 mV and STX was added to the *cis* solution. Current records were low pass filtered at 20 Hz (from Krueger *et al.* *(21)*). (B) Single K^+ channels from apical membrane vesicles prepared from cultured toad kidney cells (line A6). Membrane holding potential was +40 mV, and current traces were low pass filtered at 100 Hz (Olans and Benos, unpublished observations). (C) Single channel records of Ca^{2+}-activated, sarcoplasmic reticulum K^+ channels. At constant applied potential (+50 mV) increasing the Ca^{2+} concentration (*cis* side) results in the channel spending more time in the open state. Note that open state conductance is not modified by Ca^{2+} (from Latorre *et al.* *(25)*). In all records the arrows indicate the zero current level.

Miller (*28*) was able to incorporate a K^+-selective channel using sarcoplasmic reticulum (SR) membrane vesicles. It has become clear that the conduction characteristics of this channel are not influenced by the presence of organic solvent in the bilayer inasmuch as the channel displays comparable properties in nominally solvent-free planar bilayers (*22*). Furthermore, a K^+ conduction pathway is formed in liposomes constructed from detergent extracts of SR (*25*). More recently, Garcia and Miller (*15*), by measuring Tl^+ fluxes into SR vesicles, have concluded that these conduction sites show the same properties as the K^+ channel incorporated into planar bilayers. Specifically, they demonstrated that the ionic selectivity, conductance and channel blockade by quarternary ammonium ions are not modified by transfer of the channel from the vesicle to the planar membrane.

Figure 1 displays the bilayer setup used in the fusion method together with three examples of different channels incorporated with this method. Records of Fig. 1A correspond to sodium channels from a rat brain preparation (*21*). Inasmuch as sodium channels inactivate, the vesicles are added in the presence of batrachotoxin (BTX). BTX is known to open sodium channels (*1*). Slow sodium channel fluctuations are induced by addition of saxitoxin (STX) to the vesicle-containing side. In bilayers, STX blocks the sodium channels in a voltage-dependent manner. Records of Fig. 1B show the activity of a K^+ selective channel present in an apical membrane vesicle preparation of cultured toad kidney cells (Olans and Benos, unpublished observations). Although no clear physiological role for this epithelial channel is known at present, K^+ channels with similar conductances have been found in the apical membrane of frogskin and *Necturus* gallbladder (for a review see Benos (*3*)). Figure 1C shows current fluctuations due to the presence of a single Ca^{2+}-activated K^+ channel in a planar lipid bilayer membrane (*25*). This channel is controlled by both Ca^{2+} and by membrane potential and is found in numerous cells and tissues (*24*, *26*). This channel displays essentially the same properties in planar bilayers and in cells. In the latter case its properties have been measured with the patch-clamp technique (*26*).

2. Rationale for Channel Reconstitution

The fact that the integral membrane proteins which form ionic channels can be transferred safely and reproducibly to planar bilayers has several important consequences. First, it affords the opportunity to study physiologically relevant channels otherwise not accessible to conventional electrophysiological techniques. Prominent examples are the Ca^{2+}-activated K^+ channels contained in transverse tubule (t-tubule) membranes (*25*), the Cl^- channel from *Torpedo* electroplax (*53*) and, of course, the K^+ channel from SR (*28, 32*). In all cases, the morphology of the cell or organelle in question is such that their membranes cannot be reached by standard microelectrodes or patch-clamp pipettes. Thus, any reconstitution approach used is advantageous in that these "inaccessible" channels can be studied. Second, once the channel has been incorporated into the planar bilayer, the experimenter has easy access to either side (*i.e.*, the "external" or "cytoplasmic" face) of the channel as well as exact control of the bathing solution composition and electrical potential. Third, in many instances only one type of channel is detected in the bilayer despite the probable existence of a myriad of different channels in the native membrane. It is possible that this occurs because of the ionic conditions present during channel transfer or to a channel or vesicle "selection" process performed by the bilayer itself. For example, although a light fraction of striated muscle membranes (*35, 42*) contains both functional Ca^{2+}-activated K^+ and Na^+ channels (*35*; Moczydlowski and Miller, unpublished observations), the Ca^{2+}-activated K^+ channel can be studied without interference when the incorporation is done in a potassium medium. Similarly, only Na^+ channels are observed when the skeletal muscle vesicles contact the planar bilayer in sodium medium. Likewise, White and Miller (*53*) found a Cl^- channel while trying to incorporate the acetylcholine (ACh)-activated channel of *Torpedo* electroplax into solvent-containing bilayers. However, ACh channels are present in this vesicle preparation since, subsequently, it has been functionally reconstituted into planar bilayers made either from monolayers or with the dip method (*e.g.*, ref. *36*)*

* It is puzzling why the ACh receptor channel has resisted incorporation into solvent-containing membranes by the fusion method. Two possibilities occur to us: the planar selects only one type of vesicle, *e.g.*, inside- or right-side out, and/or solvent present in the planar bilayer (usually *n*-decane) damages the ACh receptor channel. There is some preliminary evidence suggesting that inside-out vesicles fuse more readily with solvent-containing mem-

Therefore, by either selecting the ionic composition of the bathing medium or the type and lipid composition of the planar bilayer (*i.e.*, solvent-containing *vs.* solvent-free), it is possible to detect only a single type of channel despite the fact that the native membrane may contain more than one kind of conductance pathway.

Last but not least, it is essential to have a method of assessing the health of a channel during the different stages of its biochemical purification. Functional integrity of a channel by necessity must be tested through reconstitution techniques.

3. Fusion Below the Lipid Phase Transition Temperature and Other Considerations

The ACh receptor channel has been reconstituted into solvent-free bilayer membranes by the fusion method (*6, 17*). Boheim *et al.* (*6*) were able to fuse membrane fragments rich in ACh receptor protein from *Torpedo marmorata* with a planar bilayer constructed from a synthetic mixed chain lecithin. Fusion occurs a few degrees below the liquid-solid phase transition temperature (~30°C). A clear advantage of the fusion technique employed by Boheim's group is that it avoids the use of solvent-containing bilayers and that it never exposes the channel-forming integral proteins to an air phase as is the case when using folded bilayers (*e.g.*, see refs. 40 and 48; see below). Hanke *et al.* (*6*) reported that while the presence of Ca^{2+} in the aqueous solutions or negatively-charged lipids are not a prerequisite for fusion, application of an osmotic gradient is necessary for high rates of fusion. We think that these conclusions are also true when using solvent-containing membranes. For example, we have observed that high rates of incorporation of both the t-tubule Ca^{2+}-activated K^+ channel as well as the A6 apical membrane Na^+ channel into phosphatidylethanolamine (neutral) or phosphatidylethanolamine-phosphatidylserine membranes can be obtained in virtually Ca^{2+}-free solutions, as long as an osmotic gradient is established across the planar bilayer. In solvent-containing bilayers fusion can proceed, albeit at a very slow pace, even in the absence of an osmotic gradient across the planar bilayer. However, it appears that an osmotic

branes (*12*). If this is the case, it implies that the ACh receptor channel is restricted to right-side out vesicles. To our knowledge there is no evidence damage to channels due to the presence of decane in the bilayer.

gradient is an absolute requirement for fusion when using solvent-free membranes (see ref. *22*). The effects of osmotic gradients as a requisite for fusion with planar bilayers have been discussed by Cohen *et al.* (*7*, *8*). These authors concluded that the osmotic swelling of the phospholipid vesicles adsorbed to the planar bilayer causes fusion.

When using the fusion method, incorporation proceeds either in packages of many channels (*28*, *53*) or in a one-by-one fashion (*e.g.*, ref. *25*). The simplest explanation for these phenomenon is that the membrane vesicles either contain many channels or one or no channels. In general it is convenient from the point of view of statistical analysis of channel records to have only a single channel in the membrane. Hanke and Miller (*18*) reported incorporation of a single Cl^- channel into solvent-free membranes if the *Torpedo* electroplax vesicles are sonicated for about 2 min. It is likely that sonication reduces vesicle size and hence diminishes the average number of channels per vesicle. However, the reader should be aware that Horn and Lange (*20*) presented a method for statistical data analyses of current fluctuations measured from bilayers containing more than one channel.

We would like to emphasize that we have been discussing fusion in a rather loose way. It has not been shown that the channels incorporated into solvent-containing membranes (except for mitochondrial porin (see ref. *7*)) are actually transferred from the native vesicle into the planar bilayer by a process of fusion. The evidence for fusion is only indirect, in that the experimental conditions known to increase the rate of fusion (*7*, *8*) also increase the rate of channel incorporation. The main message to be extracted here is that the optimal conditions for incorporation of ionic channels from vesicles to planar bilayers must be determined empirically for each preparation.

II. FOLDED BILAYERS

As depicted in Fig. 2A, channel incorporation into a planar bilayer can be implemented by forming lipid monolayers at an air/water interface from a suspension of vesicles (*40*, *48*, *49*). Figure 2A illustrates the case in which one of the monolayers actually is a proteolipid because the liposomes from which this monolayer is formed contain integral pore-forming protein(s). Assembly of two monolayers across a hydrophobic partition (*e.g.*, Teflon) gives as a result a lipid bilayer containing the channels. According to Schindler (*47*) the only absolute requirement for successful bilayer formation from vesicles is that the vesicle radii must be greater than 50 nm. Schindler found that with increasing vesicle size the monolayer surface pressure increases, reaching a limiting value of about 35 dynes/cm. The rate of equilibration between vesicles and monolayers appears to increase by the addition of divalent cations

Folded Bilayers

A. Monolayers formed from vesicles

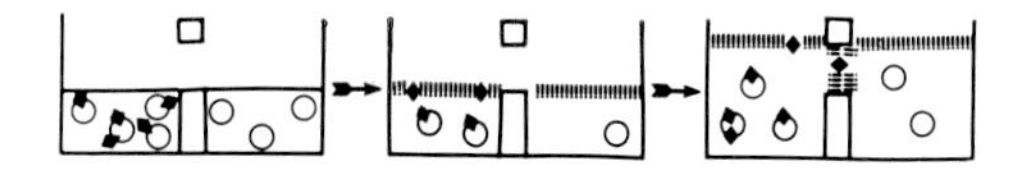

B. Monolayers formed from exogenous lipids

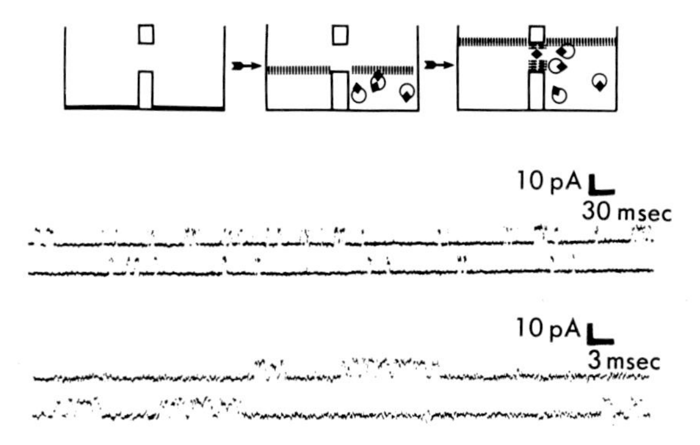

Fig. 2. Folded bilayer set-up. (A) Both compartments of a bilayer chamber are filled with a vesicle suspension to a level just below the membrane aperture. Note that the vesicles containing channel proteins are added only to one compartment. Monolayers will self-assemble at the air-water interface. Monolayers are then opposed to form a planar bilayer by raising the solution levels above the membrane aperture. In (B), monolayers can be spread using N_2-dried, purified lipids, followed by addition of channel-containing vesicles to one aqueous subphase. The planar bilayer is then formed by raising the levels of each subphase. Potassium channels obtained from lobster walking nerve axons measured in folded bilayers are also shown at different time scales at 4.5 kHz and 9 kHz filtering, respectively. Recordings were made at a holding voltage of 0 mV, 0.5 M KCl *cis* and 0.1 M KCl *trans* (Data taken from Coronado *et al.* (*13*)).

(*e.g.*, Mg^{2+}, Ca^{2+}). The primary disadvantage of this method is that the proteins are partially exposed to the air phase which may cause denaturation. In addition, the hole must be pretreated with a hydrocarbon solvent, usually hexadecane. Although minimal, these membranes may contain some solvent.

Figure 2B shows a method that is essentially a combination of that of Montal and Mueller (*37*) and Schindler and Rosenbusch (*49*). This method consists of forming the monolayer from a film of previously dried lipids. Monolayers at the air/water interface are formed by adding electrolyte solution to one compartment, and electrolyte plus vesicle-containing solution to the other (*13*). At present, it is not clear whether the channels in this case are incorporated into the bilayer as shown in Fig. 2A or by fusion (*i.e.*, Fig. 1). However, the rate of channel incorporation is larger in the presence of an osmotic gradient suggesting that at least some fusion is occurring. As in the case of the Schindler and Rosenbusch (*49*) method (Fig. 2A), pretreatment of the hole with hexadecane is an absolute prerequisite for membrane stability.

Table I shows that ACh receptor channels as well as porins have been incorporated into planar bilayers by forming monolayers from vesicles containing those channels. In this regard, the very complete and elegant work of Montal *et al.* (*36*) has established unequivocally that the purified ACh receptor channel can be successfully reconstituted in planar bilayers. However, as pointed out by Miller (*29*) a great disadvantage is the difficulty of applying agonist molecules rapidly which brings as a consequence that the ACh receptor channel dwells in its desensitized state. We think that this problem can be obviated by reconstituting the channel in bilayers made on patch clamp pipettes (see below).

The channel activity shown in the records of Fig. 2 corresponds to that of a K^+ channel from lobster axons incorporated by the method outlined in Fig. 2B. This channel displays most of the characteristics of the delayed rectifier. A detailed account of these channels' properties is given in ref. *13*.

III. DIP BILAYERS: BILAYERS MADE FROM MONOLAYERS ON PATCH-CLAMP PIPETTES

The assembly of phospholipid bilayers from monolayers on standard patch clamp pipettes is a reconstitution strategy which has been recently promulgated for the study of single ionic channels (*13*, *36*, *50*, *51*). The main advantage of this method over both the fusion and folded bilayer techniques is the small surface area of membrane used. This fact permits the study of reconstituted ionic channels with a greatly increased time and current resolution with low background noise. Electrical seals between the bilayer and the glass routinely range from 1–10 gigaohms. Pipettes either fire-polished or not have been used. Apparently, the combination of fire-polishing and coating the pipettes with Sylgard (to within 40–50 μm of the tip) is better in terms of higher resistance seals and smaller capacitative transients. This technique has the advantage of not requiring the presence of alkane solvents for bilayer formations. Thus, the channels are always contained in a lipid environment.

Figure 3 depicts the procedure used to form "dip" bilayers. A solution-filled patch pipette is first removed from a solution on which rests a phospholipid monolayer made from either native channel-containing vesicles (*51*) or exogenous lipids. At this point the lipid head groups of the monolayer contact the aqueous solution in the pipette while the hydrocarbon tails contact air. Upon reintroducing the pipette into the monolayer solution, a bilayer membrane is formed. When bilayers are formed from exogenous lipids, it is convenient to establish an osmotic gradient across the pipette tipe (vesicle-containing side hyperosmotic) to promote channel incorporation into the bilayer from channel containing membrane fragments or liposomes. Several current records of calf sarcolemma Cl^- and K^+ channels in dip bilayers are also shown in Fig. 3. Other channels that have been observed using this technique are listed in Table I.

Schuerholz and Schindler (*50*) report a slightly different procedure for forming bilayers on patch pipettes. In their method, the patch pipette is first pushed through a saline-filled hole in a Teflon plate onto which a lipid monolayer has previously been spread. This plate is

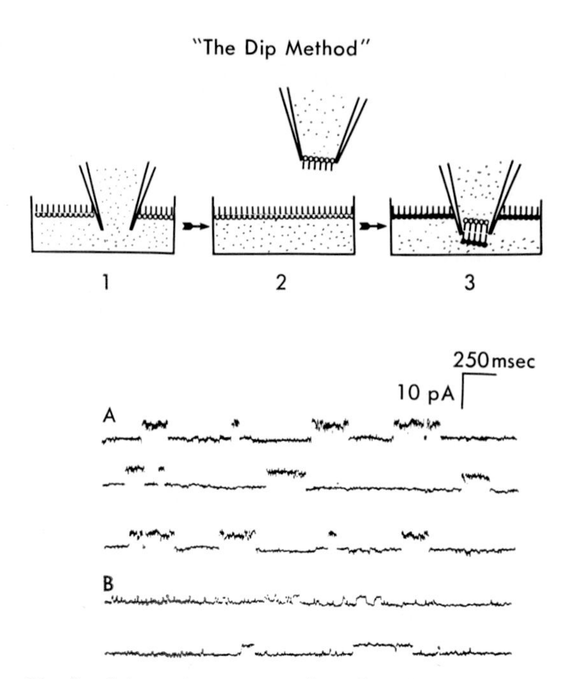

Fig. 3. Schematic representation of method used to form bilayers at the tip of patch-clamp micropipettes (the dip method). A monolayer of phospholipid is first spread at an air-water interface after immersion of saline-filled patch pipette in the solution (frame 1). The pipette is moved out of the solution taking with it a film of adsorbed phospholipid (frame 2). Reentry of the pipette into the same (or different) phospholipid monolayer solution results in the formation of a stable bilayer (frame 3). Channels can be incorporated either by manufacturing channel containing monolayers (see Fig. 2) or by adding channel-containing vesicles to the aqueous subphase (in frame 3) and allowing fusion to proceed. Recordings of single-channel currents using this method are also shown. (A) corresponds to a 70 pS chloride channel (holding potential of −50 mV) and (B) corresponds to a 35 pS potassium channel at −80 mV. Both channels were obtained from a preparation of calf cardiac sarcolemma. Records were low-pass filtered at 1 kHz and 500 Hz, respectively. Data are from Coronado and Latorre (*11*).

suspended a very short distance above a second monolayer-containing trough. As the pipette moves through the first hole it immediately contacts the monolayer resting on the lower aqueous phase forming a bilayer. These authors report electrical seals of 2–5 TΩ. Membrane channels of mitochondrial matrix protein (porin) and colicin A were studied with this technique. However, in their paper, they include a technical note stating that when bilayers were constructed by dipping into and out of a monolayer containing solutions (see Fig. 3), bilayer stability was always <1 min. They claim that stable bilayers could be formed only if solvent were included in the phospholipid solution. This difficulty with bilayer stability has not been encountered in our own or in other investigators' experience (*11*, *36*, *51*).

The use of patch pipettes to study biological ionic channels extracted, purified, and reconstituted into large (>10 μm diameter) liposomes has also been reported (*51*, *52*). In this case, a "patch" of liposomal membrane is transferred to the tip of a glass pipette by suction under direct microscopic visualization. Both the *Torpedo* electroplax Cl^- channel (*52*) and the ACh receptor channel (*51*) have been studied in this way.

IV. CONCLUSIONS

A significant development arising from the field of membrane transport in recent years is the incorporation into model membranes of ionic channels from biological membranes. This development makes amenable a precise and quantitative study of these transport systems in an environment of defined and controllable composition. The techniques which we have reviewed allow for a detailed electrical study of channels that could not otherwise have been done by conventional methods. For example, aside from accessibility, the study of channels in planar bilayers permits experimentation on the effects of membrane lipid composition on channel function. Further, we are now in a position to modify ionic channels either chemically or biochemically in a situation free from other complicating cellular events. Hence, channel structure-function relationships can now be studied on a molecular level. Furthermore, once a channel is reconstituted and once its functionality is established independent of its native membrane, an unequivocal demonstration of the actual protein components required for channel physiological activity can be achieved.

SUMMARY

Methods of ionic channel reconstitution using planar lipid bilayers are reviewed. These methods include 1) the fusion of channel-containing vesicles to preformed planar bilayers, 2) the formation of planar bilayers from monolayers (folded membranes), and 3) the construction of bilayers at the tip of patch clamp pipettes (the dip method). The advantages and disadvantages of each method are assessed.

Acknowledgments

The friendship, support, insight and wit of Dr. C. Miller is cheerfully acknowledged. We thank Ms. T. Hardenbergh for superb secretarial assistance. This work was supported by NIH Grants GM-28992, GM-31768, and AM-25886; Fondo Nacional de Investigacion, Universidad de Chile, Proyecto B 1224 8333; and by funds from the Andrew W. Mellon Foundation.

REFERENCES

1 Albuquerque, E.X. and Daly, J. (1976). In *Receptors and Recognition, Ser. B.*, vol. 1, ed. Cuatrecasas, P., pp. 297–338. London: Chapman & Hall.
2 Bayliss, W.M. (1915). *Principles of General Physiology*, London: Longmans.
3 Benos, D. (1983). *Comments Mol. Cell Biophys.* **2**, 111–128.
4 Benz, R., Janko, K., Boos, W., and Lauger, P. (1978). *Biochim. Biophys. Acta* **454**, 305–319.
5 Bernstein, J. (1902). *Pflugers Arch.* **92**, 521–562.
6 Boheim, G., Hanke, W., Barrantes, F.J., Eibj, H., Sakmann, B., Fels, G., and Maelicke, A. (1981). *Proc. Natl. Acad. Sci. U.S.* **78**, 3586–3590.
7 Cohen, F., Akabas, M., and Finkelstein, A. (1982). *Science* **217**, 458–460.
8 Cohen, F., Zimmerberg, J., and Finkelstein, A. (1980). *J. Gen. Physiol.* **75**, 251–270.
9 Colombini, M. (1979). *Nature* **279**, 643–645.
10 Coronado, R. and Latorre, R. (1982). *Nature* **298**, 849–852.
11 Coronado, R. and Latorre, R. (1983). *Biophys. J.* **43**, 231–236.
12 Coronado, R. and Williams, A. (1982). *Biophys. J.* **37**, 343a.
13 Coronado, R., Latorre, R., Mautner, H.G. (1984). *Biophys. J.* **45**, 289–299.
14 Dawson, C.M., Atwater, I., and Rojas, E. (1982). *J. Memb. Biol.* **64**, 33–43.
15 Garcia, A.M. and Miller, C. (1984). *Biophys. J.* **45**, 49–51.
16 Hamill, O.P., Marty, A., Neher, E., Sakmann, B., and Sigworth, F.J. (1981). *Pfluger's Arch.* **391**, 85–100.
17 Hanke, W., Eibl, H., and Boheim, G. (1981). *Biophys. Struct. Mech.* **7**, 131–137.
18 Hanke, W. and Miller, C. (1983). *J. Gen. Physiol.* **82**, 25–45.
19 Hodgkin, A.L. and Huxley, A.F. (1952). *J. Physiol. (Lond.)* **117**, 500–544.
20 Horn, R. and Lange, K. (1983). *Biophys. J.* **43**, 207–223.
21 Krueger, B.K., Worley, J.F., and French, R.J. (1983). *Nature* **303**, 172–175.
22 Labarca, P., Coronado, R., and Miller, C. (1980). *J. Gen. Physiol.* **76**, 397–424.
23 Latorre, R. and Alvarez, O. (1981). *Physiol. Rev.* **61**, 77–150.
24 Latorre, R. and Miller, C. (1983). *J. Memb. Biol.* **71**, 11–30.
25 Latorre, R., Vergara, C., Hidalgo, C. (1982). *Proc. Natl. Acad. Sci. U.S.* **79**, 805–809.
26 Latorre, R., Vergara, C., and Moczydlowski, E. (1984). *Cell Calcium*, **4**, 343–357.

27 Marty, A. (1983). *Pflugers Arch.* **396**, 179–181.
28 Miller, C. (1978). *J. Memb. Biol.* **40**, 1–23.
29 Miller, C. (1983). *Physiol. Rev.* **63**, 1209.
30 Miller, C., Arvan, P., Telford, J.N., and Racker, E. (1976). *J. Memb. Biol.* **30**, 271–282.
31 Miller, C. and Racker, E. (1976). *J. Memb. Biol.* **30**, 283–300.
32 Miller, C. and Rosenberg, R. (1979). *J. Gen. Physiol.* **76**, 307–424.
33 Miyazaki, S. and Igusa, Y. (1982). *Proc. Natl. Acad. Sci. U.S.* **79**, 931–935.
34 Moczydlowski, E. and Latorre, R. (1983). *Biochim. Biophys. Acta* **732**, 412–420.
35 Moczydlowski, E. and Latorre, R. (1983). *J. Gen. Physiol.* **82**, 511–542.
36 Montal, M., Labarca, P., Fredkin, D.R., Suarez-Isla, B.A., and Lindstrom, J. (1984). *Biophys. J.* **45**, 165–174.
37 Montal, M. and Mueller, P. (1972). *Proc. Natl. Acad. Sci. U.S.* **69**, 3561–3566.
38 Mueller, P. and Rudin, D.O. (1969). *Curr. Topics Bioenerg.* **3**, 157–249.
39 Neher, E. and Sakmann, B. (1976). *Nature* **260**, 799–802.
40 Nelson, N., Anholt, R., Lindstrom, J., and Montal, M. (1980). *Proc. Natl. Acad. Sci. U.S.* **77**, 3057–3061.
41 Noble, D. (1978). *The Initiation of the Heartbeat*, Oxford: Clarendon Press.
42 Rosemblatt, M., Hidalgo, C., Vergara, C., and Ikemoto, N. (1981). *J. Biol. Chem.* **252**, 5565–5574.
43 Sariban-Sohraby, S., Burg, M.B., Latorre, R., and Benos, D. (1983). *J. Gen. Physiol.* **82**, 25a–26a.
44 Sariban-Sohraby, S., Latorre, R., Burg, M., Olans, L., and Benos, D. (1984). *Nature* **308**, 80–82.
45 Shein, S.J., Kagan, B.L., and Finkelstein, A. (1978). *Nature* **276**, 159–163.
46 Shein, S.J., Colombini, M., and Finkelstein, A. (1976). *J. Memb. Biol.* **30**, 99–120.
47 Schindler, H. (1980). *FEBS Lett.* **122**, 77–79.
48 Schindler, H. and Quast, U. (1980). *Proc. Natl. Acad. Sci. U.S.* **77**, 3052–3056.
49 Schindler, H. and Rosenbusch, J.P. (1978). *Proc. Natl. Acad. Sci. U.S.* **75**, 3751–3755.
50 Schuerholz, T. and Schindler, H. (1983). *FEBS Lett.* **152**, 187–190.
51 Suarez-Isla, B.A., Wan, K., Lindstrom, J., and Montal, M. (1983). *Biochemistry* **22**, 2319–2323.
52 Tank, D.W., Miller, C., and Webb, W.W. (1982). *Proc. Natl. Acad. Sci. U.S.* **79**, 7749–7753.
53 White, M.M. and Miller, C. (1979). *J. Biol. Chem.* **254**, 10161–10166.
54 Wilmsen, U., Methfessel, C., Hanke, W., and Boheim, G. (1983). In *Physical Chemistry of Transmembrane Ion Motions*, ed. Troyanovsky, C., pp. 479–485. Amsterdam: Elsevier.
55 Young, R.C., Allen, R., and Meissner, G. (1981). *Biochim. Biophys. Acta* **640**, 409–418.

17

ION PERMEATION THROUGH SARCOPLASMIC RETICULUM MEMBRANE

KAZUKI NUNOGAKI AND MICHIKI KASAI

Department of Biophysical Engineering, Faculty of Engineering Science, Osaka University, Osaka 560, Japan

Sarcoplasmic reticulum (SR) regulates the cytoplasmic Ca^{2+} concentration of skeletal muscle cells and thereby controls muscular contraction and relaxation (*2*). In connection with these functions, the Ca^{2+} pump mechanism of the SR has been established as the cause of relaxation (*2, 4, 5, 19*). However, the molecular mechanism of Ca^{2+} release from SR is still unknown (*4*). Knowledge of movement of other ions during the uptake and release of Ca^{2+} is important to understand these functions at the molecular level.

Many researchers have measured the ionic permeability of SR membrane by various methods; however, there are discrepancies among the data reported. For example, for K^+ permeability we obtained by the light scattering method a value of 20 sec for half permeation time, $\tau_{1/2}$ (5.8×10^{-8} cm/sec for permeability coefficient, P; refs. *7* and *11*.). McKinley and Meissner reported that the K^+ permeability is greater than 10^{-6} cm/sec from measurement of membrane potential using a potential probe (*14*). Chiu and Haynes obtained a value around 10^{-6} cm/sec ($\tau_{1/2}=2$ sec) using anilinonaphtalenesulfonate (ANS) fluorescence (*1*).

Miller and his coworkers observed a channel conductance of 120 pS in 100 mM K-gluconate using vesicles incorporated into an artificial planar bilayer (*12*, *15*). This value corresponds to $\tau_{1/2}$ of 1.1 msec ($P = 1.0 \times 10^{-3}$ cm/sec) assuming that one vesicle contains only a single channel (*10*), a much higher value than those obtained by other methods (*1*, *7*, *11*, *14*). These discrepancies are mainly attributable to the difference of the method employed. The light scattering method could not follow permeation faster than water ($\tau_{1/2} = 0.1$ sec), nor could the ANS method follow permeation faster than ANS ($\tau_{1/2} > 20$ msec). It is probable that slower components of the K^+ permeation were observed by these two methods, and that the conductance measurement observed the fastest component.

Recently we applied the fluorescence quenching technique to the measurement of the rapid permeation of ions, studying the influx of Tl^+ and I^- instead of K^+ and Cl^- (*17*). Through these experiments, we determined the reason for variations in the reported data on the ionic permeability of SR membrane obtained by different methods (*10*).

I. PREPARATION OF SR MEMBRANE VESICLES

SR vesicles were isolated as microsomal fractions by the method of Weber *et al.* (*20*) with minor modifications (*8*, *21*). From rabbit dorsal and hind leg muscle, two types of SR vesicles were prepared: heavy SR (HSR) which sediments at $4{,}000 \times g$ and light SR (LSR) which sediments at $44{,}000 \times g$. LSR vesicles contain mainly calcium pump proteins and HSR vesicles have calcium binding proteins and are mostly derived from terminal cisternae through which calcium ions are released *in vivo*.

II. PERMEABILITY MEASUREMENT BY TRACER METHOD

The simplest method to determine permeability is the tracer method. Figure 1 shows an example of the permeability measurements for some ions; the radioactivity of the remaining ions in the vesicles was normalized by the total radioactivity. From this figure, the following facts can be seen: 1) efflux rates are different from ion to ion, and 2) the values extrapolated to zero time are different. Permeability for various

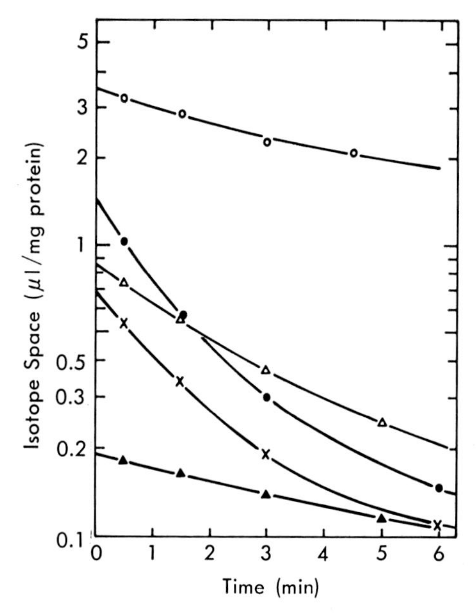

Fig. 1. Permeability measurement by tracer method. SR vesicles were incubated in 0.2 osm of various salts containing radioactive tracers, 5 mM Tris-maleate (pH 6.5) and 10 mg protein/ml at 0°C for overnight. Efflux of various ions from SR vesicles followed after 100-fold dilution in the same solution used for the incubation without the labeled ions. Ordinate represents the relative radioactivity remaining in the vesicles, R/R_0C, where R is the radioactivity in the vesicles, R_0 that before filtration, and C the SR concentration during the incubation. (○) ^{14}C-choline$^+$ (choline-Cl), (●) $^{35}SO_4^{2-}$(K_2SO_4), (△) $^{22}Na^+$ (NaCl), (×) $^{42}K^+$ (KCl), (▲) $^{36}Cl^-$(NaCl).

ions has been determined from the slope of these curves, however it is incorrect because the second point was not taken into account. Since the amount of the permeant in SR vesicles was normalized by the total radioactivity, these values correspond to the apparent intravesicular space (referred to as the apparent volume). The apparent volumes for these ions were smaller than the intravesicular water space determined by inulin exclusion volume (about 3–5 μl/mg protein, ref. *6*).

The following explanations are possible for the difference in the apparent volume. 1) The permeability for the ion was so small that the ion was not equilibrated during the overnight incubation and the amount of the tracer incorporated was small. 2) In contrast, the permeability was so high that some ions had already flowed out during the filtration. 3) Some ions can bind to the inside of the membrane and give a high apparent volume. 4) Even at the equilibrium condition, the intravesicular ion concentration was different from the extravesicular one due to the Donnan effect (*6*).

In choline$^+$, the efflux curve represents the complete time-course since the apparent volume was close to the intravesicular water space determined by the inulin exclusion volume, no specific binding was observed, and the time-course of choline influx determined by other methods was consistent with the curve in Fig. 1. Factor 3) operates in the case of divalent cations such as Ca^{2+} (*9*); case 1) applies to sucrose (*6, 9*); with other ions, the main effect is attributable to case 2). In other words, the fast components of the efflux could not be followed by the tracer assay. Accordingly, permeability coefficients determined from the slope of these curves are only those of the slow components.

III. PERMEABILITY MEASUREMENT BY LIGHT SCATTERING METHOD

The tracer method cannot follow the efflux faster than a few seconds. One method of following faster flux is to measure the changes in scat-

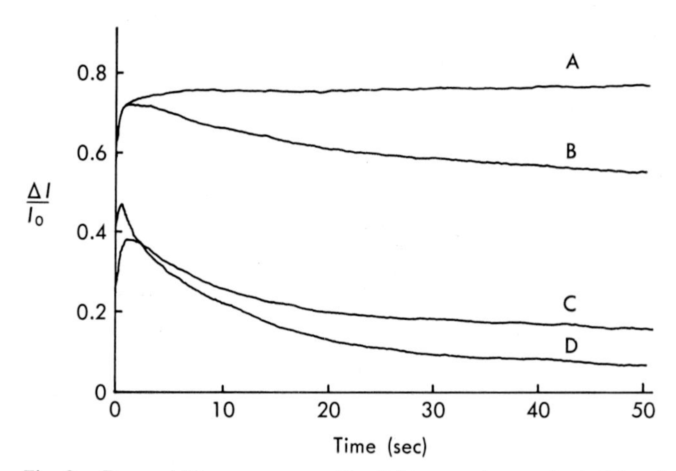

Fig. 2. Permeability measurement by light scattering method. SR vesicles were incubated in 5 mM Tris-maleate and 0.2 mg protein/ml for 1 hr at 23°C. The suspension was rapidly mixed with solutions containing various salts of 0.2 osm, 5 mM Tris-maleate (pH 6.5) and the change in scattered light intensity was followed. (A) sucrose, (B) choline-Cl, (C) KCl, (D) glycerol.

tered light intensity caused by the osmotic volume change of the vesicles (*7, 11*); Fig. 2 shows an example. When the osmolarity of the solution was increased, the scattered light intensity rapidly increased and then decreased. The rapid increase in the intensity is due to the shrinkage of the vesicles caused by water efflux; this rate gives water permeability. The subsequent decrease in the intensity is due to swelling of the vesicles caused by an influx of water accompanied by an influx of solutes. From its rate, the permeability for the solutes can be determined. For example, the half permeation time was 0.1 sec for water, and those for KCl, choline-Cl, glycerol and sucrose were 10 sec, 270 sec, 1.8 sec and 2,000 sec, respectively (Permeability for choline depends on the presence of μM Ca^{2+}, see ref. *21*). In salt permeation, ions are considered to move as a pair according to the Nernst equation (*11*), then the permeation rate of the less permeable ion becomes rate determining. In the case of KCl, when valinomycin was added to the solution, the permeation rate increased (*11*), showing that the permeability for K^+ was smaller than that for Cl^-. Analysis gave a permeation time of 20 sec for K^+ and 0.4 sec for Cl^-. By this method, the permeability for various ions has been determined (*11*).

Though the extrapolated value of the falling phase to zero time corresponds to the volume of the vesicles at the most shrunken state and is therefore expected to be equal for all solutes, the value differed among solute species. For example, the extrapolated value for KCl was about half of that for choline-Cl. This discrepancy can be explained by the existence of vesicles with larger permeability for some salts, *e.g.*, KCl, than water, because such vesicles do not change their volume with an osmolarity jump. Thus, our preparation turned out to contain at least two types of vesicles: one with channels highly permeable to K^+ and Cl^- and the others without such channels. As far as the component observed here is concerned, we can say that the permeability for K^+ is smaller than that for Cl^-.

For solutes such as choline-Cl or glucose, all components of the permeation can be followed by this technique. These curves agree well with those of the tracer experiments. When we measured the permeability for rapidly permeating neutral molecules such as glycerol, the maximum increment of the scattered light intensity was high; they are considered to permeate through all vesicles by solvation in the lipid phase. This suggests that the mechanisms of the permeation of ions and neutral molecules are different; ions pass the channels.

IV. MEASUREMENT OF INFLUX OF Tl^+ AND I^- BY FLUORESCENCE QUENCHING METHOD

In order to follow the faster permeation, we developed a fluorescence quenching method (*17*) after Moore and Raftery (*16*). They measured the Tl^+ influx into acetylcholine receptor rich membrane vesicles using anilinonaphthalenetrisulfonate (ANTS) as a fluorophore. We used a less permeable dye, pyrenetetrasulfonate (PTS). SR vesicles were equilibrated with PTS and free PTS outside the vesicles was removed by a Sephadex column. The eluent was rapidly mixed with a Tl^+ solution and the subsequent quenching of the PTS fluorescence due to

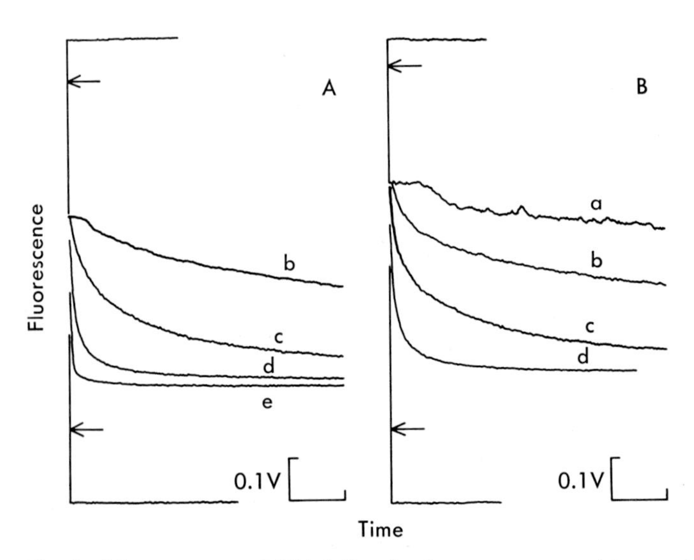

Fig. 3. Measurements of Tl^+ influx by fluorescence quenching method. SR vesicles containing 1 mM PTS eluted with 0.1 M KNO_3 solution were rapidly mixed with 0.1 M $TlNO_3$ solution at time zero. The subsequent decrease in PTS fluorescence was monitored in several time ranges: The scale bars represent (a) 2 msec, (b) 20 msec, (c) 0.2 sec, (d) 2 sec, (e) 20 sec. The lowest traces are the background of the solvent and the upper traces are the total fluorescence intensity measured by mixing with 0.1 M KNO_3 instead of 0.1 M $TlNO_3$. Upper arrows show the level of fluorescence inside the vesicles at zero time. Lower arrows show the level of fluorescence fully quenched. (A) measurement by the double mixing cell, (B) measurement by the single mixing cell.

Tl^+ influx was followed by a stopped flow apparatus, as shown in Fig. 3. It is known that Tl^+ is able to pass through various kinds of K^+ channels (*13*). Tl^+ permeates SR membrane as fast as K^+ when measured by the light scattering method. The quenching of fluorescence obeyed the Stern-Volmer law (*3*). A little leakage of PTS was measured by the millipore filtration method and the instantaneous quenching of extravesicular PTS was estimated. Thus, in Fig. 3, the quenching curve should begin from the level indicated by an arrow on each trace. The initial decreasing phase could not be recorded because of dead time of the stopped flow apparatus. With a single mixing cell, the dead time was about 1 msec and with a double mixing cell, it was about 6 msec. For each specimen, we obtained two sets of data with these two mixing cells.

The curve in Fig. 3 could not be explained by simple exponential influx. We made an analysis under the following assumption: 1) the sample was a mixture of several kinds of vesicles with different ionic permeabilities, 2) the quencher influx into each type of vesicle was exponential with a representative permeation time. In formula, the fluorescence intensity was given as:

$$F(t)=F_{\circ}\sum_{i}\frac{C_i}{1+KQ_i(t)} \tag{1}$$

where C_i is the fraction of the component i and $\sum C_i=1$. $Q_i(t)$ is the concentration of the quencher in vesicles of component i, and is expressed by

$$Q_i(t)=Q_{\circ}\left(1-\exp\left(-\frac{t}{\tau_i}\right)\right) \tag{2}$$

where τ_i is the permeation time for the component i.

Analysis was made graphically with a microcomputer by the fitting and subtraction of each component in order from the slowest one. The data of the double mixing cell were analyzed first and the slower components were determined. Then the data of the single mixing cell were analyzed using the parameters for slower components. Some results are shown in Fig. 4. Three main components were obtained: typically, 1.5 msec 40%, 50 msec 16%, 1.4 sec 30%. In Fig. 3, the initial phase could not be recorded, but if we extrapolate the fastest

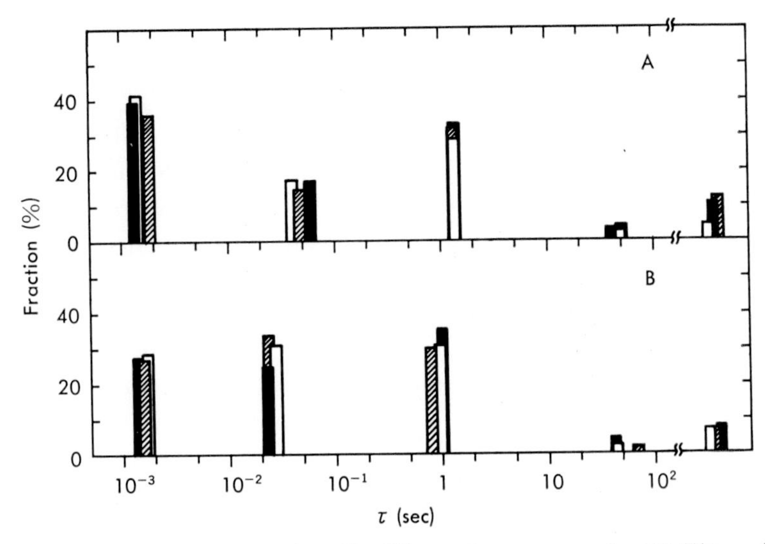

Fig. 4. Distribution of vesicles with different time constants for (A) Tl^{+}- and (B) I^{-}-influx. The time constant and the fractional amplitude of each component were obtained from experiments similar to Fig. 3 using LSR. Different symbols represent different preparations.

component to zero time, we can explain the total change in the fluorescence intensity. Thus there is no faster permeation than those observed here.

Similar experiments were made with HSR. The results showed that the permeation time for each component was approximately the same as that of LSR, but the fractional amplitude of the fastest component was larger (about 70%). This indicates that HSR preparation involves many vesicles having the highly permeable channel (see the following section)

The anion permeability was also studied in the same way as that for cation using I^{-}. KI permeates as fast as KCl into SR vesicles when measured by the light scattering method. Three main components were obtained: 1.6 msec 28%, 24 msec 30%, 0.9 sec 32%. For HSR, an increase of the fastest component was observed as in the case of Tl^{+} influx.

Now we compare these results with those obtained by other methods. In the light scattering method, the 3rd and 4th components are considered to be measured. The fact that the permeation time of the

3rd component is longer for Tl^+ than for I^- is consistent with the results by light scattering experiments that anion is more permeable than cation. The facts that the sum of the 4th and 5th components is rather small for each ion and the sum for I^- is smaller than that for Tl^+ are also consistent with the results of tracer experiments where the apparent volume was much smaller than the intravesicular water space and the apparent volume for Cl^- was smaller than that for K^+.

V. COMPARISON OF FLUORESCENCE QUENCHING MEASUREMENTS WITH CONDUCTANCE MEASUREMENTS

According to Miller's method, Sokabe and his coworkers have measured single channel currents through SR membrane prepared in our laboratory (*10*, *18*). They found the cation channel reported by Miller and an anion channel. The cation channel has one open state and one closed state, and single channel conductance of 160 pS for K^+ and 80 pS for Tl^+. The anion channel has several open states and the single channel conductance of its most conducting state is 200 pS for Cl^- and 180 pS for I^- (Sokabe and Tanifuji; private communication).

Here we compare these channel conductances with the fastest component of our quencher influx experiments. The time constant of ion influx into a vesicle of volume v involving N channels of single channel permeability p is given by

$$\tau = \frac{v}{Np} \tag{3}$$

The single channel conductance of the channel measured under the symmetrical ionic condition of concentration C is

$$\gamma = pC\frac{z^2F^2}{RT} \tag{4}$$

Combining Eqs. 3 and 4, the time constant for a vesicle with one channel (τ_{cal}) can be calculated as 2.44 msec for Tl^+ and 1.08 msec for I^-. Dividing τ_{cal} by τ_{exp} gives a mean number of open channels per vesicle having the channel (n_o^*): 1.6 for a cation channel and 0.7 for an anion channel. Since about 1/7 of the cation channels are open at zero membrane potential (*12*), an average of about 11 such channels are expected to exist in a vesicle. This value agrees with the conductance measurement (Sokabe; private communication). For an anion channel, the value 0.7 for n_o* is also consistent with the observation by conductance measurement that this channel fluctuates among several lower conducting states and only one channel is primarily involved in a vesicle (Sokabe; private communication).

If we assume that our preparation is homogeneous in membrane nature and the cation channels are distributed randomly over all vesicles, their distribution should obey the Poisson statistics. The obtained number 1.6 for n_o* corresponds to the mean channel number of all vesicles (n_o) of 1.1. The fraction of vesicles having the open channel is calculated to be 63%†. Since this value is larger than the experimental result of 40%, it is suggested that the SR vesicle preparation is inhomogeneous and is composed of two classes of vesicles which are different in membrane nature.

REFERENCES

1 Chiu, V.C.K. and Haynes, D.H. (1980). *J. Membrane Biol.* **56**, 203–218.
2 Ebashi, S. and Endo, M. (1968). *Prog. Biophys. Mol. Biol.* **18**, 123–183.
3 Eftink, M.R. and Ghiron, C.A. (1976). *J. Phys. Chem.* **80**, 486–493.
4 Endo, M. (1977). *Physiol. Rev.* **57**, 71–108.
5 Hasselbach, W. (1964). *Prog. Biophys. Mol. Biol.* **14**, 167–222.
6 Kasai, M. (1980). *J. Biochem.* **88**, 1081–1085.
7 Kasai, M., Kanemasa, T., and Fukumoto, S. (1979). *J. Membrane Biol.* **51**, 311–324.
8 Kasai, M. and Miyamoto, H. (1976). *J. Biochem.* **79**, 1053–1066.
9 Kasai, M. and Miyamoto, H. (1976). *J. Biochem.* **79**, 1067–1076.
10 Kasai, M., Nunogaki, K., Nagasaki, K., Tanifuji, M., and Sokabe, M. (1984). In *Sarcoplasmic Reticulum: Structure and Function*, ed. Tonomura, Y. and Fleischer, S., New York: Academic Press, in press.
11 Kometani, T. and Kasai, M. (1978). *J. Membrane Biol.* **41**, 295–308.
12 Labarca, P., Coronado, R., and Miller, C. (1980). *J. Gen. Physiol.* **76**, 397–424.
13 Latorre, R. and Miller, C. (1983). *J. Membrane Biol.* **71**, 11–30.
14 McKinley, D. and Meissner, G. (1978). *ibid.* **44**, 159–186.
15 Miller, C. (1978). *ibid* **40**, 1–23.
16 Moore, H.-P.H. and Raftery, M.A. (1980). *Proc. Natl. Acad. Sci. U.S.* **77**, 4509–4513.
17 Nunogaki, K. and Kasai, M. (1982). In *Abstract of the 20th Annual Meeting of the Biophys. Soc. Japan*, p. 70, Osaka.
18 Sokabe, M., Kometani, T., Tanifuji, M., and Kasai, M. (1982). *ibid.*, p. 72, Osaka.

† According to our simulation which takes into account the size distribution of SR vesicles, n_o* is about 2 and the volume fraction of vesicles with open channels is 92%.

19 Tonomura, Y. (1972). In *Muscle Proteins, Muscle Contraction and Cation Transport.*, pp. 305–356, Tokyo: Japan Sci. Soc. Press.
20 Weber, A., Herz, R., and Reiss, I. (1966). *Biochem. Z.* **345**, 329–369.
21 Yamamoto, N. and Kasai, M. (1982). *J. Biochem.* **92**, 465–475.

18

MODULATION OF A SPECIFIC K+ CHANNEL IN *APLYSIA* NEURONS BY SEROTONIN AND cAMP

JOSEPH CAMARDO

Center for Neurobiology & Behavior, New York State Psychiatric Institute, and College of Physicians and Surgeons, Columbia University, New York 10032, U.S.A.

Many different biological signals are communicated by means of the intracellular second messenger, cyclic AMP (cAMP). This transmembrane signaling mechanism begins at the extracellular surface when a hormone or neurotransmitter binds to the membrane receptor of a cell and activates adenylate cyclase, an enzyme that increases the level of cAMP inside the cell. A subsequent intracellular cascade of cAMP-dependent processes activates protein kinase, an enzyme that phosphorylates specific proteins and modifies their function. This alters cellular functions in which these proteins play a role (*1, 2*). One result of this is that a localized membrane event (transmitter binding) can initiate widespread changes in the physiology of the cell (many different proteins are phosphorylated), so the transmembrane signal can be amplified. The second messenger system is quite prominent in the nervous system (*3*), and it is not surprising that one of the targets of cAMP-dependent phosphorylation in excitable cells is the membrane currents themselves.

Many of the cAMP-mediated actions of transmitters in various excitable membranes fall into the class of slow postsynaptic potentials (slow PSP) (*4*). A distinctive feature of the slow PSP is that the transmitter modulates membrane currents which contribute to the passive properties of the cells and to the resting potential and the action potential. The slow PSP can therefore alter the electrical behavior of the neuron by changing the integrative properties (space and time constants) and the threshold and configuration of the action potential. Since the action potential itself triggers yet another second messenger, the Ca^{2+} ion, modulation of these currents will also modify aspects of cell function coupled to the influx of Ca^{2+} (*5*). Moreover, these changes are relatively long-lasting, usually from seconds to minutes, in contrast to the fast PSP such as that elicited by acetylcholine at the neuromuscular junction.

This chapter describes a slow PSP which serotonin (5-hydroxytryptamine, 5-HT) elicits in an identified group of sensory neurons in the marine mollusc *Aplysia californica* by means of the second messenger, cAMP. First, whole cell voltage-clamp experiments which show that serotonin modulates a specific K^+ current will be briefly summarized (*6, 7*). Second, single-channel recording experiments which elucidate the action of serotonin on single ion channel proteins will be presented in detail (*8*). Other aspects of the role of the slow PSP in the physiology of these cells can be found in a recent review by Kandel and Schwartz (*9*).

I. SEROTONIN AND cAMP DECREASE A SPECIFIC K^+ CURRENT IN *APLYSIA* SENSORY NEURONS

Serotonin, or stimulation of the nerve pathway which is thought to release serotonin, elicits a slow, decreased conductance, excitatory postsynaptic potential (slow EPSP) in sensory neurons. This depolarizes the cell, decreases the conductance, and delays repolarization of the action potential (Fig. 1A and 1B). In the voltage-clamped sensory cell (Fig. 1C), serotonin decreases the total outward current, both instantaneous and steady state current, at all potentials equal to or depolarized from the resting potential. Evidence that this effect is mediated by cAMP comes from both physiologic and biochemical experiments. First, both the slow PSP in the unclamped cell and the same decrease in membrane current can be exactly reproduced by intracellular injection of cAMP.

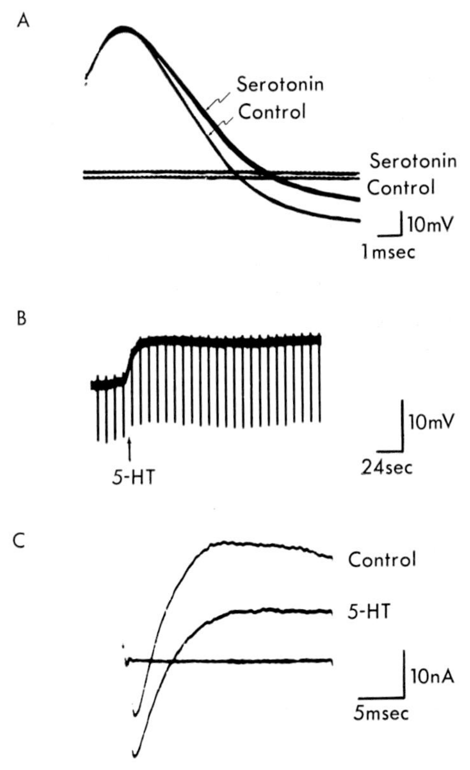

Fig. 1. Serotonin elicits a slow, decreased conductance EPSP in sensory neurons of *Aplysia*. (A) and (B). Intracellular recordings from sensory neurons illustrate the three main effects of the serotonin induced slow EPSP: depolarization, delayed repolarization of the action potential, and increased membrane resistance. In 1A, action potentials in control and serotonin-treated cells were evoked by intracellular stimulation, and the rising phases and peaks were aligned. The repolarization phase is delayed, and the after-hyperpolarization is decreased following exposure to serotonin. The cell also depolarizes by a few mV (bottom traces). In 1B, hyperpolarizing constant current pulses are injected into the cell at resting potential to measure the resistance. After serotonin, the cell depolarizes, and the membrane resistance increases. This is a direct effect of serotonin; if the cell is depolarized to the same potential by injecting depolarizing current, the resistance decreases. (C). Voltage-clamp records show the effect of 5-HT on the membrane currents of the sensory cell. The cell was clamped to −50 mV, and the voltage was stepped to 0 mV for 25 msec. Only the current record is shown; the horizontal line is the holding current at −50 mV. The voltage step elicits an early inward current, and a later outward current; serotonin decreases net outward current at all points during the step. This results in an apparent increase in the peak inward current, and a decrease in the peak outward current. The steady-state outward current is also reduced (not shown). These effects occur by means of a decrease by serotonin of a steady-state K^+ current, which is activated even at the resting potential, and contributes outward current throughout the voltage step.

Second, biochemical studies show that serotonin selectively elevates cAMP in sensory neurons (*10*).

The ionic basis of the slow PSP was examined in experiments by Klein and Kandel (*7*). Three findings indicate that the slow PSP results from a decrease in outward potassium current. First, serotonin decreased a current that is sensitive to the extracellular concentration of K^+. Second, the action of serotonin was eliminated by pharmacologic blockade of K^+ channels. Third, the decrease in current was blocked by replacing intracellular K^+ with the nonpermeant cation, cesium (Cs^+). Additional experiments by Klein, Camardo and Kandel (*6*) showed that the membrane of the sensory cells of *Aplysia* contains three K^+ currents which have also been described in other molluscan neurons (*11*): (1) the early K^+ current (I_A); (2) the delayed K^+ current (I_K); and (3) the Ca^{2+}-activated K^+ current (I_C). Each of these currents can be characterized by its voltage dependence, its susceptibility to pharmacologic blocking agents, and, in the case of the Ca^{2+}-activated current, by its dependence on intracellular Ca^{2+}. Klein *et al.* found that serotonin did not act on these three K^+ currents, and postulated the existence of a fourth K^+ current, specifically sensitive to serotonin and cAMP. They called the new current S current, by analogy with M current, a muscarine-sensitive current discovered in sympathetic neurons of the frog (*12*).

Voltage-clamp studies have provided some information about the serotonin-sensitive current. It is activated at the resting potential, contributes to repolarization of the action potential, is moderately voltage-dependent, remains activated with steady-state depolarization, and is not dependent on intracellular Ca^{2+}. Its decrease can therefore account for the effects of serotonin and cAMP on the unclamped sensory cell.

II. SINGLE CHANNEL RECORDING CAN BE USED TO STUDY THE ACTION OF SEROTONIN

Implicit in the conclusions drawn from voltage-clamp studies is the existence of a specific K^+ channel, which is modified in some way by serotonin and cAMP. This hypothesis can be tested directly with single-channel patch-clamp recording. This technique (*13, 14*) allows one to measure current through single ion channels in their normal membrane environment, as the channels change from conducting (open) to non-conducting (closed) states. Current through a single channel is in the pico-amp range, and the resolution of pico-amps is achieved by means of a high resistance (gigaohm, 10^9 ohms) seal between the electrode and a small patch of cell membrane (see Fig. 2).

The relationship between single channel and total membrane current is quite simple. Assume that a channel obeys the following scheme:

$$\text{closed} \underset{\alpha}{\overset{\beta}{\rightleftharpoons}} \text{open}$$

where α and β represent first-order rate constants for closing and opening of the channel, respectively. The average current at any instant (I), through either the patch membrane or the whole-cell membrane, is the mean current through all the channels, and is given by the equation:

$$I = n \cdot p \cdot i \tag{1}$$

where n is the number of functional channels, p is the probability of finding a channel open, and i is the current through a single open channel. The probability is a function of the opening and closing rates, and is given by $p = \beta/(\beta+\alpha)$. Serotonin could decrease total K^+ current by changing n, p, or i. The transmitter could change the current through a single channel by decreasing the conductance of individual channels (i). Transmitter could decrease p in two ways. First, it could shorten the mean open time of the channel by increasing the closing rate constant α, so that the channel closes more rapidly after opening. Second, it could decrease the probability of channel opening by decreasing the opening rate constant β, so that channels open less frequently, but once open would stay open for their usual time. Finally, a transmitter could prevent a particular channel from contributing current at all, and therefore decrease n, the number of functional channels in the membrane.

III. SEROTONIN AND cAMP CLOSE SINGLE POTASSIUM CHANNELS

We used the experimental design shown in Fig. 2 to address two questions. First, is there a serotonin-sensitive K^+ channel? Second, which feature of the channel behavior is altered by serotonin? Single chan-

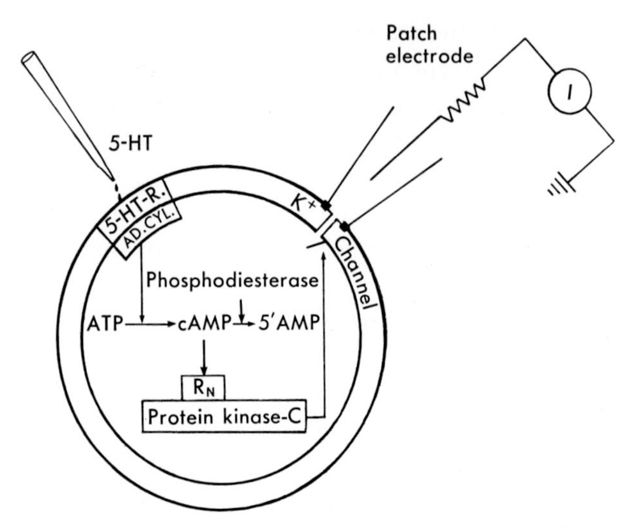

Fig. 2. Schematic illustration of the experimental single channel recording protocol and the cAMP-dependent model for the action of serotonin. A high-resistance (gigaohm) seal is obtained between the extracellular fire-polished patch electrode and the cell membrane by applying gentle suction to the electrode. Single-channel currents in the patch of membrane under the electrode can then be measured and the potential of the patch can be changed while the cell is held at rest. All current generated by the patch must enter the electrode; the gigaohm seal prevents leakage of current to the bath. This is the cell-attached recording configuration. If the patch electrode is withdrawn from the cell, the seal remains intact and single-channel currents can be recorded from the cell-free patch of membrane with the cytoplasmic surface of the membrane exposed to the bathing solution. 5-HT-R., 5-HT-receptor; AD.CYL., adenylate cyclase; R_N, regulatory subunit of protein kinase. Protein kinase represents the catalytic subunit of this enzyme.

nels were recorded from a patch of membrane on a sensory cell exposed to trypsin to promote seal formation, and the cell resistance was monitored with an intracellular electrode. Serotonin is added to the bath. Since the gigaohm seal prevents transmitter from entering the patch pipette from the bath, any effect of serotonin on channels in the patch must occur by means of an intracellular second messenger. This experiment is therefore also a test of the second messenger hypothesis. After the giga-seal is obtained, steady-state depolarization is applied to the patch membrane, while the cell itself is held at resting potential. This procedure maximizes the appearance of the steady state S current in the patch, by inactivating the strongly voltage-dependent early and delayed K^+ currents (I_A & I_K), and minimizing Ca^{2+} influx, to prevent activation of the Ca^{2+}-dependent current.

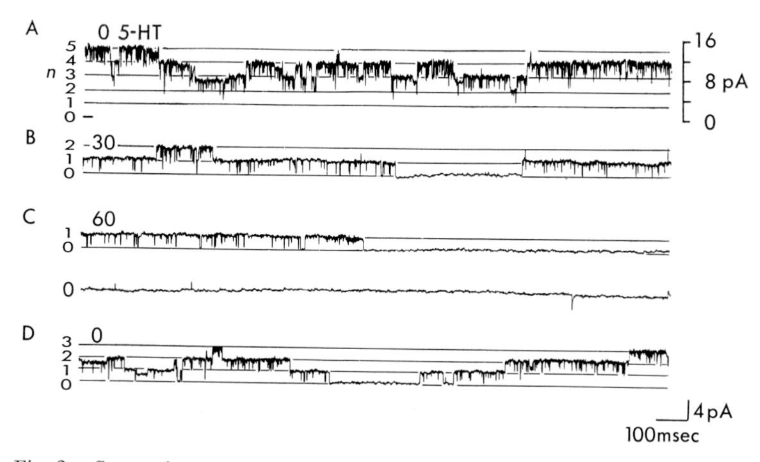

Fig. 3. Serotonin causes prolonged closure of a steady-state K^+ channel. These records were obtained from a sensory cell, using the method illustrated in Fig. 2. Representative samples of patch current were taken during various parts of the experiment. The number of open channels is shown to the left of each trace (n), and the total current is shown to the right of the top trace. Outward current is upward in these records. The cell-resting potential is −40 mV, and the steady-state potential of the patch electrode is also −40 mV. The transmembrane potential across the patch is therefore 0 mV, *i.e.*, $V_{trans} = V_{in} - V$ electrode. In trace A, a total of five channels fluctuate randomly between five levels of current. Note that current never decreases to zero, and for the entire time of the record, which is representative of a much longer control period, at least three, and usually four channels are open for most of the time. Trace B shows that channel opening is decreased by serotonin, at a concentration of 30 μM. Current in this trace is never greater than two open channels (at least three channels are always closed), and all five channels are closed for several hundred msec; this never occurred in the absence of serotonin. In trace C, the one remaining open channel is closed by increasing the concentration of serotonin to 60 μM. This channel remains closed for several minutes (lower part of record C). Trace D shows that partial recovery of channel opening occurs following washout of serotonin.

As is illustrated in Fig. 3, in the control patch held at 0 mV, five outward channels of uniform size open and close randomly, maintaining a certain average patch current. After addition of serotonin, and accompanying a decrease in whole-cell conductance (not shown), the average patch current decreased in a stepwise fashion, due to the closure of three of the five active channels. Additional serotonin closed the remaining channels and the total patch current fell to zero. At least three channels resumed opening when the serotonin was washed out. Exactly the same effect occurs when cAMP is injected into the cell (Fig. 4). Serotonin and cAMP do not produce channels of smaller

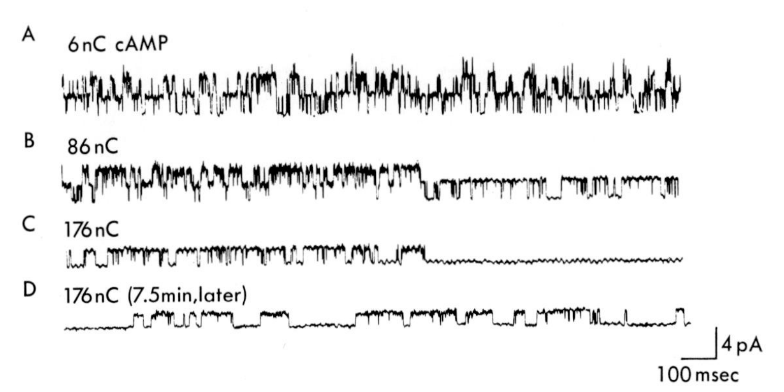

Fig. 4. Effect of intracellular injection of cAMP on single-channel current. After an on-cell patch was obtained and a control record taken, the cell was impaled with a micro-electrode filled with 1 M cAMP (Sigma) and cAMP was injected into cell by hyperpolarizing current pulses (0.5–3.5 nA). Total charge injected into cell (nanocoulombs) is indicated on top of each trace. (A). Current record obtained soon after impalement of the cell with cAMP electrode. (B, C). Current records after subsequent injection periods. During the records shown in C, the remaining active channel closed. The cAMP electrode was withdrawn from the cell and no channel openings were observed for 5 min. Channel activity then partially recovered, as shown in trace D. Channel size is smaller in C and D because the cell hyperpolarized slightly during the experiment.

amplitude (i is not decreased), nor channels which close more rapidly (α is unchanged). The best interpretation of the stepwise decrease in current is that some channels no longer open, and the effect of 5-HT and cAMP is therefore equivalent to a decrease in n in Eq. 1. Of course, it is also possible that β is reduced to zero, so that the opening rate is essentially zero. Although mechanistically different, for practical purposes this is the same as a decrease in n.

These experiments therefore answer the two questions we posed. The serotonin-cAMP cascade culminates in a decrease of K^+ current through a *specific channel*, as indicated by the whole-cell voltage-clamp data, and this is accomplished by prolonged closure of the channel.

IV. PROPERTIES OF THE SEROTONIN SENSITIVE CHANNEL

We have studied the biophysical properties of the S channel in cell-attached patches, and also extended our analysis using the cell-free patch. In this configuration the membrane patch remains adherent to the electrode when it is withdrawn from the cell and channel activity

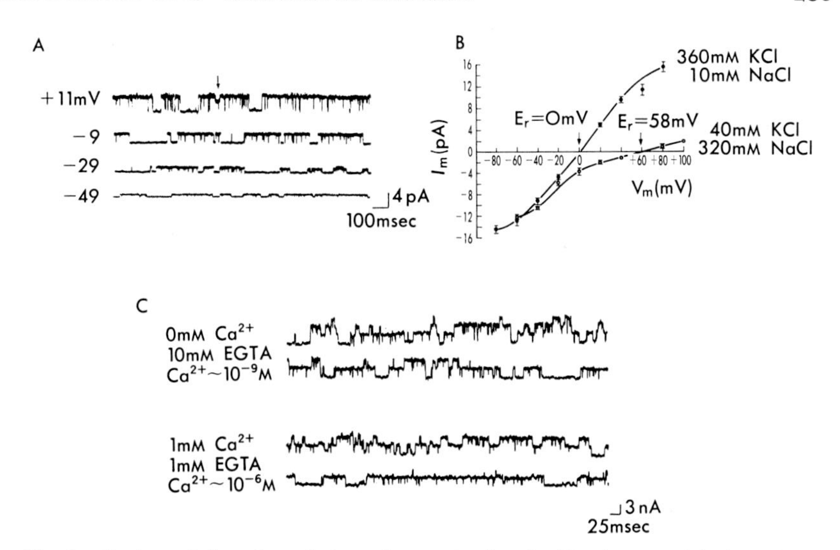

Fig. 5. S channel is voltage-independent, selective for K^+ ions, and independent of intracellular Ca^{2+}. (A). The S channel remains open over a wide range of membrane potential in the cell-attached patch configuration. The patch membrane potential was altered by changing the potential inside the recording electrode while the intracellular resting potential (−39 mV) was constant. The *trans*-patch potential is to the left of each trace. Both the bath and patch electrode contain artificial seawater with 10 mM KCl. Channel openings appear as step increases in outward current (outward current plotted in upward direction). The current fluctuates between two levels corresponding to the fully closed and fully open channel and channel size increases with depolarization away from the K^+ equilibrium potential. The channel is open at the resting potential and below (−49 mV) and remains open with steady-state depolarization. The arrow marks a partial closing. (B). Reversal potential for S channel current follows the Nernst equation for K^+. This experiment was done with the cell-free patch, with 360 mM KCl in the electrode pipette. The two curves were done with different concentrations of K^+ in the bath. (●) The current-voltage curve obtained in symmetric 360 mM KCl shows the reversal potential to be 0 mV; in this solution the conductance is greater than 200 pico-siemens and the IV curve is linear throughout most of its range. The inner surface of the patch was then exposed to 40 mM KCl. For a pure K^+ current, the reversal potential, from the Nernst equation, is +55 mV. (○) The reversal potential in this asymmetric solution is about +58 mV, the IV curve rectifies, and the slope conductance at 0 mV is approximately 120 pico-siemens. The predicted and observed reversal potentials are very close, and indicate that the channel is selective for K^+ ions. (C). Opening of the S channel is independent of intracellular Ca^{2+}. In this experiment with a cell-free inside-out patch (cytoplasmic surface faces the bath), the K^+ concentrations were asymmetric; 10 mM KCl in the electrode (external surface), 360 mM KCl in the bath (internal surface), and K^+ current flows outward. The Ca^{2+} concentration in the bath was changed using a Ca^{2+}-EGTA buffer. The current records show that the channel remains open in approximately 1 nM Ca^{2+} (top), and its opening is not increased when the Ca^{2+} is raised to approximately 1 μm (bottom).

can still be recorded. The cytoplasmic surface of the membrane is now exposed to an artificial bathing solution, which can be changed to examine the effects of ions, such as Ca^{2+}, or solutions of different composition on channel current. Our experiments show that the S channel has the properties predicted for it from voltage-clamp experiments.

The S channel is open at the resting potential, and remains open with steadystate depolarization; it is therefore noninactivating (Fig. 5A). It is open at a wide range of potentials, and its opening is only moderately affected by voltage so it is aptly described as a background channel. The channel has a slope conductance of 55 pico-siemens at 0 mV in the cell-attached patch, and rectifies in the outward direction due to the asymmetry of the K^+ concentrations across the membrane (*15*). Both the conductance and the rectification are functions of the K^+ ion concentration; rectification disappears when the K^+ concentrations on both sides of the patch are equal, and the conductance increases with higher K^+ concentration (Fig. 5B).

The S channel is highly selective for K^+ ions. Figure 5B shows the effect on the reversal potential of a change in K^+ concentration, from 360 mM on each side (where the reversal potential is zero) to 360 mM on the outside of the membrane (inside the electrode) and 40 mM on the inside of the membrane. The Nernst equation for a pure K^+ current predicts a change in reversal potential of 55 mV for this nine-fold change in K^+ concentration. We observed a change of 58 mV, which is within the range of experimental error. An intermediate concentration of K^+ also gave the predicted reversal potential.

Channel opening is independent of intracellular Ca^{2+} (Fig. 5C). The probability of the channel being open is unchanged when the Ca^{2+} concentration at the cytoplasmic surface of the membrane is increased from 1 nM to 1 μM. This means that, on the cell, the channel is likely to be open at all Ca^{2+} concentrations. By contrast, the Ca^{2+}-activated K^+ channel of the sensory neurons, similar to that described in other cells (16~18), is profoundly affected by intracellular Ca^{2+}. It is closed at Ca^{2+} concentrations below 10^{-6} M, where the S channel is open. This Ca^{2+}-dependent K^+ channel underlies the current elicited by injection of Ca^{2+} into *Aplysia* neurons (*6*, *19*); it has a slightly smaller conductance than the S channel.

These experiments provide good evidence that this channel is indeed the protein which underlies S current. It satisfies the requirements predicted from voltage-clamp data and its closure will account for the action of serotonin on the sensory cell.

V. POSSIBLE ROLE OF PROTEIN PHOSPHORYLATION IN CHANNEL CLOSURE

Our data support the hypothesis that serotonin causes prolonged closure of a specific K^+ channel in *Aplysia* sensory neurons by means of the intracellular second messenger cAMP. Since all of the known actions of cAMP occur by means of activation of cAMP-dependent protein kinase, our data therefore strongly implicate protein phosphorylation as a key element of channel closure. Further support for this comes from two other experiments. First, the slow EPSP can be elicited by injection of purified catalytic subunit of cAMP-dependent protein kinase (*20*), and, second, the effect of serotonin can be blocked by protein kinase inhibitor (*21*). However, the substrate for phosphorylation, and how phosphorylation affects the channel, remain unknown.

There are many ways in which alterations in protein function by phosphorylation could modulate the S channel. One can imagine that the channel opening is regulated by an ion pump in the cell, which itself is modified by protein phosphorylation. A cytoplasmic protein which regulates channel gating could be altered by phosphorylation, as could a membrane-associated channel regulatory protein. The channel could be completely removed from the membrane, perhaps by an endocytosis-like mechanism. A most appealing and simple hypothesis is that the channel itself is phosphorylated, but there is no direct evidence for this. We have recently begun experiments using the cell-free patch to examine these various possibilities, but our data do not support any firm conclusion. Nevertheless, it is clear that the transmembrane signal which increases intracellular cAMP can exert profound changes in specific membrane ion channel proteins, which are strategically placed such that their modification dramatically changes the electrophysiological behavior of the neurons. The molecular basis of these protein modifications is a subject for future study.

SUMMARY

The neurotransmitter serotonin elicits a slow, decreased conductance EPSP in sensory neurons in the mollusc *Aplysia californica.* Serotonin depolarizes the cells, increases the input impedance, and delays repolarization of the action potential, all by means of a decrease in potassium current. Serotonin also elicits a specific biochemical effect in these cells: the receptor is coupled to adenylate cyclase, and therefore the transmitter elevates intracellular cAMP. The electrophysiologic effects of serotonin on these neurons can be produced by intracellular injection of cAMP, and the serotonin-induced EPSP is therefore thought to occur *via* this second messenger, and ultimately to require protein phosphorylation.

Our recent work has examined the effect of serotonin on membrane K^+ current and single K^+ channels. We find that serotonin decreases a specific membrane K^+ current, and exerts its effects exclusively on this current, which we call serotonin-sensitive, or S current. Three other K^+ currents in these cells, the fast current (I_A), delayed rectification (I_K), and the calcium-dependent current (I_C) are unaffected by serotonin. We have also studied the K^+ channel which carries this serotonin-sensitive current, using single channel patch clamp recording. These cells contain a K^+ channel which in all respects fits the properties predicted for the serotonin-sensitive current from voltage clamp data. This channel is highly selective for K^+, it is moderately voltage-sensitive, it remains open at the resting potential and with steady-state depolarization, and it is insensitive to calcium. Both serotonin and cAMP induce long periods of channel closure which have the effect of decreasing the effective number of open channels in the membrane at any instant. At the single-channel level, this closure accounts for the decrease in K^+ current which we see with the voltage-clamp technique.

REFERENCES

1 Cohen, P. (1980). In *Molecular Aspects of Cellular Regulation*, vol. 1, ed. Cohen, P., pp. 1 and 255. Amsterdam: Elsevier/North-Holland.

2 Rosen, O. and Krebs, E. (ed.) (1981). *Protein Phosphorylation*, Cold Spring Harbor Conferences on Cell Proliferation, vol. 8. New York: Cold Spring Harbor Laboratory.

3 Greengard, P. (1980). *Harvey Lecture* **75**, 277–331.
4 Kehoe, J. and Marty, A. (1980). *Annu. Rev. Biophys. Bioeng.* **9**, 437–465.
5 Klein, M., Shapiro, E., and Kandel, E.R. (1980). *J. Exp. Biol.* **89**, 117–157.
6 Klein, M. Camardo, J.S., and Kandel, E.R. (1982). *Proc. Natl. Acad. Sci. U.S.* **79**, 5713–5717.
7 Klein, M. and Kandel, E.R. (1980). *Proc. Natl. Acad. Sci. U.S.* **77**, 6912–6916.
8 Siegelbaum, S.A., Camardo, J.S., and Kandel, E.R. (1982). *Nature* **299**, 413–417.
9 Kandel, E.R. and Schwartz, J.H. (1982). *Science* **218**, 433–443.
10 Bernier, L., Castellucci, V.F., Kandel, E.R., and Schwartz, J.H. (1982). *J. Neurosci.* **2**, 1682–1691.
11 Adams, D.J., Smith, S.J., and Thompson, S.H. (1980). *Annu. Rev. Neurosci.* **3**, 141–167.
12 Brown, D.A. and Adams, P.R. (1980). *Nature* **283**, 672–674.
13 Neher, E., Sakmann, B., and Steinbach, J.H. (1978). *Pflugers Arch. Eur. J. Physiol.* **375**, 219–228.
14 Hamill, O., Marty, A., Neher, E., Sakmann, B., and Sigworth, F.J. (1981). *Pflugers Arch. Eur. J. Physiol.* **391**, 85–100.
15 Hodgkin, A.L. and Katz, B. (1949). *J. Physiol. (Lond.)* **108**, 37–77.
16 Marty, A. (1981). *Nature* **291**, 497–500.
17 Pallotta, B.S., Magleby, K.L., and Barrett, J.V. (1981). *Nature* **293**, 471–474.
18 Adams, P.R., Constanti, A., Brown, D.A., and Clark, R.B. (1982). *Nature* **296**, 746–749.
19 Meech, R.W. (1972). *Comp. Biochem. Physiol.* **42A**, 493–499.
20 Castellucci, V.F., Kandel, E.R., Schwartz, J.H., Wilson, F.D., Nairn, A.C., and Greengard, P. (1980). *Proc. Natl. Acad. Sci. U.S.* **77**, 7492–7496.
21 Castellucci, V.F., Nairn, A., Greengard, P., Schwartz, J.H., and Kandel, E.R. (1982). *J. Neurosci.* **2**, 1673–1681.

19

ATP-DEPENDENT UPTAKE OF GLUTAMATE INTO SYNAPTIC VESICLES

TETSUFUMI UEDA

Mental Health Research Institute, Departments of Psychiatry and Pharmacology, The University of Michigan, Ann Arbor, Michigan 48109, U.S.A.

L-Glutamate has long been implicated as a major excitatory neurotransmitter in the mammalian central nervous system. In the early 1960's, Curtis, Watkins, Krnjević and colleagues demonstrated (*10*, *33*) that this dicarboxylic amino acid has a potent, rapid, excitatory action on most of the central neurons, which raised the possibility of a transmitter role for glutamate (for reviews, see refs. *11*, *34*). It has been difficult, however, to establish the notion that glutamate functions as a neurotransmitter in the central nervous system. Besides its ubiquitous occurrence throughout the biological system and the large diversity of central neurons affected, it is involved in a variety of other functions such as energy and nitrogen metabolism, incorporation into proteins and peptides, serving as a precursor for the inhibitory neurotransmitter γ-aminobutyric acid, and the regulation of osmotic balance (for reviews, see ref. *26*). Nonetheless, there has been increasing evidence in recent years to support the proposed transmitter role of glutamate. Various studies

The abbreviations used are: IgG, immunoglobulin G; GABA, γ-aminobutyric acid; AMP-PCP, adenylyl-(β,γ-methylene)-diphosphate; FCCP, carbonyl cyanide *p*-trifluoromethoxyphenylhydrazone; and DCCD, *N,N'*-dicyclohexylcarbodiimide.

now point to the existence of specific postsynaptic receptors (for reviews, see refs. *8*, *29*, *35*, *38*, *46*, *61*) which are presumably coupled to sodium and/or calcium channels (for reviews, see refs. *44*, *51*; see also *43*). Equally important evidence is that glutamate is released from certain nerve terminals in a calcium-dependent manner upon their depolarization (for reviews, see refs. *1*, *8*).

However, it is not known whether glutamate is released directly from the synaptic vesicle into the synaptic cleft. If the calcium-dependent release is mediated by a process which involves synaptic vesicles, loading the vesicles with glutamate would be an important step in glutamate transmission; uptake into the vesicle would be the first step to determine the fate of glutamate as a neurotransmitter. Contrary to this premise, currently available evidence indicates no significant enrichment of glutamate in isolated synaptic vesicles (*32*), nor was there a clear demonstration of the uptake of glutamate into isolated synaptic vesicles (*12*, *45*), and De Belleroche and Bradford proposed that glutamate is released directly from the cytoplasm of nerve terminals (*13*). We have, however, recently provided evidence that glutamate is specifically taken up into certain highly purified synaptic vesicles in an ATP- and temperature-dependent, but sodium-independent, manner (*40*). This supports the notion that synaptic vesicles may be involved in glutamate synaptic transmission. The purification and some of the properties of Protein I-containing synaptic vesicles, in particular, the ATP-dependent glutamate uptake, will be presented here. Since we have purified these vesicles using highly-purified antibodies to Protein I (recently referred to as Synapsin I), a neuronal, synaptic vesicle-specific protein, I have also included background information on Protein I and its antibodies.

I. PROPERTIES OF PROTEIN I

Protein I was discovered (*55*) as an endogenous substrate for cAMP-dependent protein kinase in the membrane fractions of synaptic origin, in the course of investigation into the molecular mechanism underlying the synaptic effects of those monoamine neurotransmitters known to activate adenylate cyclase (for review, see refs. *20*, *21*, *59*). Protein I is now purified to apparent homogeneity and is known to consist of highly similar polypeptides, Protein Ia (Mr=86,000) and Protein Ib (Mr=80,000), both of which have unique physicochemical properties (*56*); they are basic (pI=10.3 and 10.2, respectively), rich in proline and glycine, and made up of a collagenase-sensitive, elongated region and a collagenase-resistant, globular region (Mr=48,000) (*60*). Protein I also serves as a substrate for Ca^{2+}/calmodulin-dependent protein kinase (*21, 30, 36, 50*), and for carboxyl methylase (*2*). cAMP-dependent protein kinase phosphorylates a serine residue located in the collagenase-resistant domain (*24, 56*), whereas Ca^{2+}/calmodulin-dependent protein kinase phosphorylates multiple sites, two serine residues in the collagenase sensitive region, and one in the collagenase-resistant region (*24*). A portion of the collagenase-resistant region is crucial for the attachment of Protein I to the membrane (*25, 58*).

Protein I is specific to nervous tissue (*55*), present in many parts of the nervous system (*4, 14, 18*), and highly localized on synaptic vesicles in the nerve terminal (*4, 15, 57*). Calcium influx into synaptosomes (*36*) or adenylate cyclase activating neurotransmitters such as serotonin and dopamine (*17, 42*) leads to an increase in the phosphorylation of Protein I, presumably on synaptic vesicles. The immunocytochemical studies have indicated that the distribution of Protein I in the brain is not in parallel with that of catecholamine-containing nerve terminals (*4*). In fact, the widespread distribution of Protein I did not permit proposing a specific association of Protein I-containing vesicles with a neurotransmitter. The question remained as to whether Protein I is present on all or many types of synaptic vesicles, or on a particular population of vesicles with a specific neurotransmitter.

II. PURIFICATION AND PROPERTIES OF ANTI-PROTEIN I ANTIBODIES

The approach we took in an effort to gain insights into the above question was to first isolate Protein I-containing synaptic vesicles by using antibodies to Protein I and then characterize them with respect to the type(s) of neurotransmitter(s) with which they are associated. In the course of this investigation, we realized that it would be essential to purify a specific anti-Protein I IgG, because of significant nonspecific binding of normal immunoglobulin G to synaptic vesicles and synaptic membranes. Therefore, an affinity column was made by conjugating purified Protein I to Affi-Gel 10 agarose beads, to which anti-Protein I antiserum was then applied. Nonspecifically bound materials were desorbed with 2 M NaCl, and specifically bound antibodies were eluted with 0.1 M citric acid (pH 2.5). Normal IgG was purified from preimmune or nonimmune rabbit serum by affinity chromatography on Protein A-conjugated sepharose.

The purified antibody preparation contained essentially two polypeptides, which corresponded to the heavy and light chains of IgG (*39*). The anti-Protein I IgG inhibited specifically the phosphorylation of Protein I either in a synaptic vesicle fraction or in a crude brain homogenate; it did not affect the phosphorylation of other substrates for cAMP-dependent protein kinase, calcium-dependent kinases or cAMP/Ca^{2+}-independent kinases (*39*). The anti-Protein I IgG blocked both the cAMP-dependent and calcium-dependent phosphorylation of Protein I. These results indicated that the anti-Protein I IgG is a substrate-directed, not kinase-directed, specific inhibitor of Protein I phosphorylation and would be useful for the investigation of the function of Protein I.

III. PURIFICATION OF PROTEIN I-CONTAINING SYNAPTIC VESICLES

A highly purified synaptic vesicle fraction (0.4 M sucrose layer) was first prepared from the purified synaptosomes derived from fresh bovine cortex essentially by the same procedure as described previously (*57*). Aliquots of the purified vesicle fraction were treated on ice for 1 hr, with (a) anti-Protein I IgG, (b) preimmune IgG, and (c) a control buffer in which the IgG's had been dissolved (*40*). Unbound IgG's were separated from synaptic vesicles by centrifugation at $200{,}000 \times g$ for 35 min; the pelleted vesicles were washed three times by suspending in 0.32 M sucrose/1 mM $NaHCO_3$ and centrifuging as above. At the end of the third wash, substantial precipitation was observed in the sample which had been treated with the anti-Protein I IgG; there was very little precipitation in the sample treated with the preimmune IgG or control buffer. The samples were centrifuged at $5{,}900 \times g$ for 10 min to pellet the precipitated vesicles; the pellets were washed and suspended in 0.32 M sucrose/1 mM $NaHCO_3$. Approximately 23% and 1.2% of protein in the original vesicle fraction were recovered in the anti-Protein I IgG- and preimmune IgG-precipitated fractions, respectively.

IV. PURITY AND PROTEIN COMPOSITION OF THE IMMUNO-PRECIPITATED SYNAPTIC VESICLES

The morphological purity of anti-Protein I IgG-precipitated synaptic vesicles is shown in Fig. 1. The immunoprecipitated synaptic vesicle fraction is largely free of contamination from other intracellular organelles and plasma membranes, and contains a virtually homogeneous population of spherical vesicles with an average diameter of around 350 Å. As judged by the morphology, the purity of this vesicle preparation was estimated to be at least 90%. In accord with this observation is biochemical evidence (*40*) which has indicated that the immunoprecipitated vesicles contain only small amounts of non-vesicular

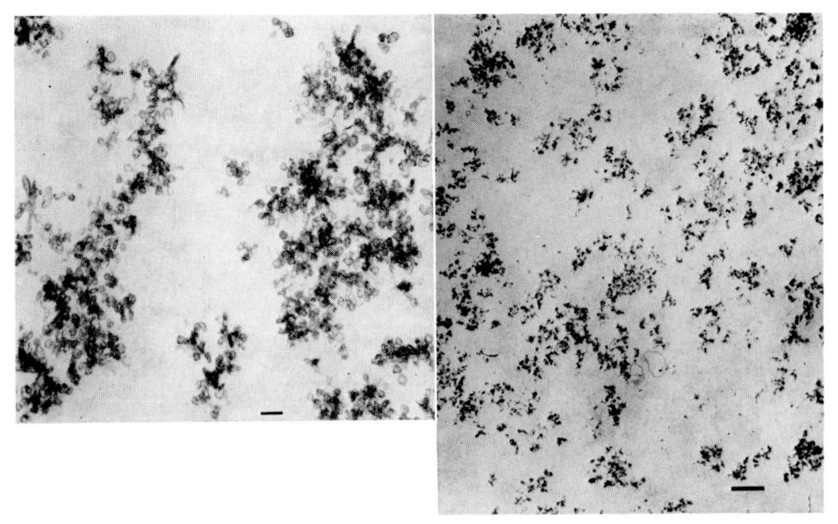

Fig. 1. Electron micrograph of anti-Protein I IgG-precipitated synaptic vesicles. The anti-Protein I IgG-precipitated synaptic vesicles were prepared from the synaptic vesicle fraction in the 0.4 M sucrose layer, followed by immunoprecipitation with affinity purified anti-Protein I IgG. The immunoprecipitated vesicle fraction contains predominantly spherical synaptic vesicles with average diameters of about 350 Å. Left, low magnification; the bar represents 1 μm. Right, high magnification; the bar represents 0.1 μm. Taken from Naito and Ueda (*40*).

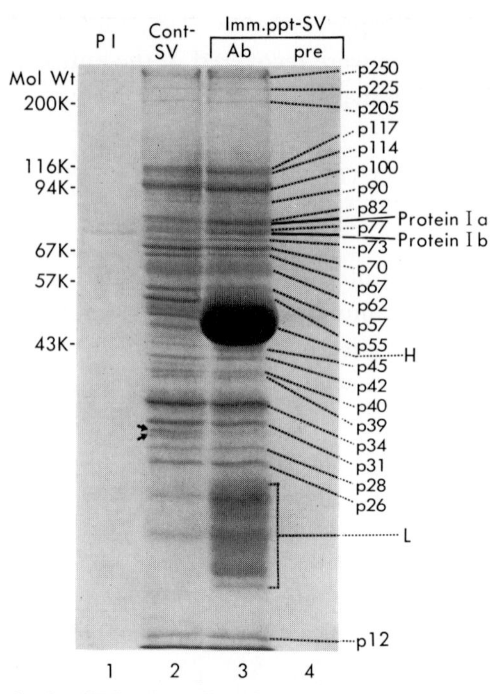

Fig. 2. SDS-polyacrylamide gel electrophoresis of synaptic vesicles before and after precipitation with anti-Protein I IgG. The synaptic vesicle fraction (2; 55 μg), which had been treated with control buffer; anti-Protein I IgG-precipitated vesicles (3; 55 μg); and preimmune IgG-precipitated vesicles (4; derived from the same amount of vesicles as used to yield lane 3) were subjected to SDS-polyacrylamide gel electrophoresis, and the gel was stained with Coomassie blue. PI, Protein I; Cont-SV, control synaptic vesicle; Imm.ppt-SV, immunoglobulin-precipitated synaptic vesicle. 200K represents 200,000, for example. Ab, antibody-precipitated vesicles; Pre, preimmune IgG-precipitated vesicles. Taken from Naito and Ueda (*40*).

marker enzymes such as cytochrome *c* oxidase, a mitochondrial marker, cytochrome *c* reductase, a microsomal marker, and (Na-K)-ATPase, a plasma membrane marker. Although the specific activities of these enzymes were already relatively low in the synaptic vesicle fraction (0.4 M sucrose layer) prepared by the differential and sucrose density gradient centrifugations, evidence indicates that the immunoprecipitation with the anti-Protein I IgG further removes significant amounts of mitochondrial, microsomal, and plasma membranous contaminants.

In contrast to the reduction in the specific activities of the nonvesicular marker enzymes, the specific contents of Protein I and possibly the 100,000-dalton protein were increased in the immunoprecipitated vesicles (Fig. 2). Concomitantly, those proteins indicated by arrows and possibly the 67,000-dalton protein were removed by the immunoprecipitation. Preimmune IgG, used as a control, caused very little precipitation. These results suggest that the anti-Protein I IgG-induced precipitation is highly specific and results in further purification of Protein I-containing synaptic vesicles.

Protein I (Ia and Ib combined) is a prominent protein in the immunoprecipitated vesicles. The other prominent proteins have minimal molecular weights of 100,000, 70,000, and 34,000. In addition, there are a number of minor proteins as labeled in the figure. The protein labeled p55, which appears to be a minor protein, probably due to the interference by the heavy chain of IgG, could be tubulin, which has been shown to be a prominent component in other synaptic vesicle preparations (*5, 62*). The proteins labeled p205, p117 (and/or p114), p57, p45, p42, p31, and p26 in our preparation could correspond to the 200 k-, 120 k-, 58 k-, 46 k-, 42 k-, 31 k-, and 27 k-dalton proteins, respectively, in the preparation of Zisapel and Zurgil (*62*). The similarity in molecular weight between the 200 k-dalton protein and myosin and the 42 k-dalton protein and actin (*62*) suggests that the anti-Protein I IgG precipitated vesicles may also contain myosin and actin. Protein I, p100, and p70, all prominent components in our preparation, are missing in the preparation of Zisapel and Zurgil (*62*). It is likely that Protein I has been removed from the synaptic vesicle during the purification, which included a gel-filtration step in the presence of 0.15 M NaCl, a condition under which Protein I is known to be easily extracted (*25, 56*).

V. ATP-DEPENDENT SPECIFIC UPTAKE OF GLUTAMATE

In an effort to define the types of neurotransmitters associated with Protein I-containing synaptic vesicles, we have examined a variety of putative neurotransmitters for their potential influx into the anti-Protein I IgG-precipitated vesicles in the absence and presence of ATP. ATP-dependency was used as an important criterion for the vesicular uptake because of the well established ATP requirement for catecholamine uptake into chromaffin granules (*28, 31*). As shown in Fig. 3,

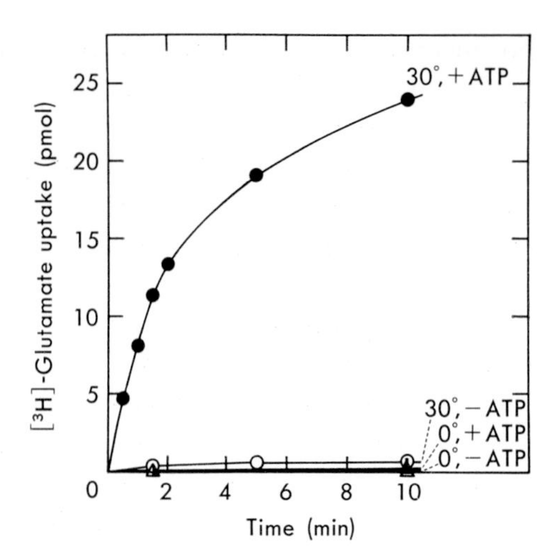

Fig. 3. Time course of the glutamate uptake into Protein I-associated synaptic vesicles. The synaptic vesicles (20 μg of protein) which had been precipitated with anti-Protein I IgG were incubated with 50 μM tritiated L-glutamate for various times indicated at 0°C (△,▲) and 30°C (○,●) in the presence (▲,●) and absence (△,○) of 2 mM ATP, and the amount of glutamate retained in the vesicles was determined by the filtration method, as described (*18*). Taken from Naito and Ueda (*40*).

TABLE I
ATP-Dependent Uptake of Various Amino Acid Neurotransmitters into Pre-immunoprecipitated, Immunoprecipitated and Non-immunoprecipitated Synaptic Vesicle Fractions

Neurotransmitter	Control buffer-treated vesicles (%)	Antibody-precipitated vesicles (%)	Antibody-nonprecipitated vesicles (%)
L-Glutamate	100	100	100
L-Aspartate	7.48±1.25 (*n*=6)	0.89±0.22 (*n*=6)*	9.17±1.67 (*n*=4)
GABA	2.87±0.49 (*n*=4)	1.43±0.45 (*n*=4)**	3.13±0.4 (*n*=3)
Glycine	0.60±0.06 (*n*=3)	0.06±0.06 (*n*=3)***	0.89±0.23 (*n*=2)

Various vesicle fractions (20 μg) were incubated for neurotransmitter uptake at 30° for 1.5 min in the absence or presence of ATP, in a mixture containing 50 μM tritiated neurotransmitter, 4 mM $MgSO_4$, 5 mM Tris-HCl (pH 7.4), and 0.32 M sucrose. The uptake activity relative to that of glutamate was expressed as percentage in each experiment. Values given are mean±S.E.M. from *n* experiments. *$p<0.0005$: **$p<0.05$: ***$p<0.002$ (by two-tailed *t*-test) compared with control buffer-treated synaptic vesicles. Modified from Naito and Ueda (*40*).

the anti-Protein I IgG-precipitated synaptic vesicles were found to take up glutamate in a highly ATP- and temperature-dependent manner. In contrast to the large stimulation by ATP, sodium had little effect; the sodium-dependent glutamate uptake was only 6.5% of the ATP-dependent uptake (*40*). This represents the first clear demonstration of the ATP-dependent uptake of glutamate into synaptic vesicles.

The ATP-dependent vesicular uptake is highly specific for L-glutamate among the putative amino acid neurotransmitters tested (Table I). Thus, in the antibody-precipitated vesicle fraction the aspartate uptake is only about 1% of the glutamate uptake; the GABA uptake is 1–2%, and the glycine uptake is 0.1% or less. All of these relative uptake activities were reduced compared to those observed in the vesicle fraction untreated with the antibody or in the fraction which had been treated with the antibody but not immunoprecipitated. According to preliminary experiments (*41*), the ATP-dependent relative uptake of acetylcholine was also reduced by the immunoprecipitation; it was not detected in the antibody-precipitated fraction, although 0.8% and 1.3% of glutamate uptake were observed in the control buffer-treated fraction and the antibody-treated but non-precipitated fraction, respectively. Since the antibody had no direct effect on the uptake of these transmitters (data not shown), it is likely that the immunoprecipitation has removed those synaptic vesicles which accumulate aspartate, GABA, glycine, and acetylcholine, and that these vesicles are distinct from those which take up glutamate, at least with respect to the transmitter transport system. In support of this view, the ATP-dependent uptake of glutamate is not inhibited by the glutamate analog glutamine, aspartate, or GABA (*40*). Monoamine neurotransmitters such as dopamine, norepinephrine, and serotonin showed, under the assay conditions used, no ATP-stimulated uptake into the vesicle fraction before or after the immunoprecipitation (*41*). These results indicate that Protein I-containing synaptic vesicles are endowed with an ATP-dependent uptake system which is highly specific for glutamate, and suggest that glutamate is concentrated *in situ* in certain synaptic vesicles. This is in accord with the recent immunocytochemical evidence indicating that glutamate is enriched in certain synaptic vesicles such as those in the mossy fiber terminals in the hippocampus, where glutamate is considered to be a neurotransmitter (*54*).

The evidence discussed above also suggests that the apparent failure of the vesicular uptake of glutamate, as reported by De Belleroche and Bradford (*12*) and Rassin (*45*), could be due to the absence of ATP from the incubation media used. It is also likely that the lack of significant enrichment of glutamate in the isolated vesicles (*32*) has resulted from a substantial leakage of glutamate from the vesicles which could have occurred during their preparation (*41*).

It should be pointed out, however, that despite the reduction in the relative uptake activities for aspartate, GABA, glycine, and acetylcholine with respect to glutamate, and the substantial increase in the specific content of Protein I (Fig. 2), there was little increase in the specific uptake activity for glutamate (*40*). This suggests that those synaptic vesicles able to take up glutamate constitute the majority of the vesicles present in the 0.4 M sucrose fraction, and raises the possibility that some of the glutamatergic vesicles (the ones which are immunoprecipitated) contain a larger amount of Protein I than do others. It is not known whether the latter occurs *in vivo*. Should it occur, the quantitative difference in Protein I content in the vesicles might reflect different stages during the life cycle of the synaptic vesicles.

The ATP-dependent vesicular uptake system for glutamate is distinct from the sodium-dependent glutamate uptake system known to be present in neural plasma membranes, including those of nerve terminals, astrocytes, cerebellar, Purkinje and granule cells (*19*, *37*, *49*). A clear difference lies in substrate specificity; the ATP-dependent glutamate uptake system hardly recognizes aspartate, whereas the sodium-dependent uptake system does not distinguish between glutamate and aspartate. Another difference is in the affinity for glutamate; K_m of the ATP-dependent uptake system has been determined to be 1.5–2.0 mM (*40a*), whereas that of the sodium-dependent uptake has been reported to be 2–40 μM (*19*, *37*, *49*). The occurrence of the low affinity glutamate uptake system in the intracellular synaptic vesicle membrane, as opposed to the high affinity uptake system in the plasma membrane, is within reason, since the intracellular concentration of glutamate is much higher than the extracellular concentration.

The specific uptake of glutamate into Protein I-containing vesicles is consistent with immunocytochemical evidence (*4*, *14*) that Protein I is not distributed in parallel with catecholamine-containing nerve endings; instead it is highly enriched in certain distinct brain regions having numerous synapses, where glutamate is probably a strong candidate for major neurotransmitter, such as in the molecular layer in the cerebellum (*7*, *22*, *47*) and the mossy fiber layer in the hippocampus (*9*,

48, *53*, *54*), but it is scarce in if not totally absent from the soma of cerebellar Purkinje cells and hippocampal pyramidal cells, which are known to be innervated by GABAergic neurons (*11*, *52*, *54*). Moreover, the degeneration of cerebellar granule cells, which are considered to be glutamatergic neurons, but not of GABAergic Purkinje cells, has been shown to result in a substantial reduction in the cerebellar content of Protein I (*16*). These observations, together with the evidence for the specific uptake of glutamate, support the notion that Protein I may be associated with glutamatergic synaptic vesicles. I suggest that a specific vesicular uptake system plays an important role in determining the type of neurotransmitters released, in particular, commonly available amino acids such as glutamate, aspartate, and glycine.

VI. MECHANISM OF THE VESICULAR UPTAKE OF GLUTAMATE

ATP hydrolysis is required for the ATP-dependent glutamate uptake into the synaptic vesicles (*40*). Thus, there was very little uptake in the absence of ATP or Mg^{2+}. Moreover, ADP and the β, γ-methylene analog of ATP, AMP-PCP, which is resistant to enzymatic hydrolysis, failed to support the glutamate uptake. GTP was far less effective than ATP. Calcium alone or in combination with magnesium had no effect (*40a*). These results suggest that a Mg-ATPase is involved in the ATP-dependent glutamate. This is supported by the observation (*41*) that the glutamate uptake was diminished by DCCD and trimethyltin, agents known to inhibit the Mg-ATPase coupled to catecholamine uptake into chromaffin granules (*3*, *6*, *27*); DCCD (80 μM) and trimethyltin (25 μM) inhibited the glutamate uptake by 87% and 99%, respectively.

The question now arises as to how the ATP hydrolysis is coupled to the glutamate uptake. In the case of catecholamine uptake into chromaffin granules, there is good evidence that electrochemical gradients, namely membrane potentials and transmembrane pH gradients, generated by a Mg-ATPase in the granule membrane, provide the driving force for the catecholamine influx (*23*, *27*, *28*). The Mg-ATPase translocates protons upon ATP hydrolysis from the exterior to the interior side of the membrane. This results in the generation of a membrane potential, positive inside, which allows the catecholamine influx to occur. In the presence of permeant anions such as chloride in the exterior medium, however, the membrane potential induces anion flux, which results in the reduction of the membrane potential, accompanied by an increase in the proton concentration inside, thereby increasing the pH gradient across the membrane. The pH gradient thus formed also causes the catecholamine uptake. Although either a membrane potential or a pH gradient alone can induce catecholamine influx, both are required for optimal catecholamine accumulation to occur.

Certain agents such as FCCP and thiocyanate have been shown to reduce the membrane potential and thereby inhibit the catecholamine uptake (*23, 27, 28*). Other agents such as nigericin (in the presence of potassium) and ammonia have been shown to inhibit the catecholamine uptake by dissipating the pH gradient across the chromaffin granule (*27*). We have examined the effects of these specific agents on the ATP-dependent uptake of glutamate into the synaptic vesicles. In a medium containing 0.25 M sucrose and 5 mM Tris-maleate (pH 7.4), a condition under which presumably a membrane potential predominates due to the absence of permeant anions, FCCP (2 μM) and thiocyanate (10 mM) were found to inhibit the glutamate uptake by 81% and 86%, respectively (*41*). In a medium containing 0.11 M KCl and 5 mM Tris-HCl, in which presumably the membrane potential is diminished and a pH gradient is formed, nigericin (7 μM) and ammonia (30 mM) caused 74% and 76% of inhibition, respectively (*41*). These results suggest that

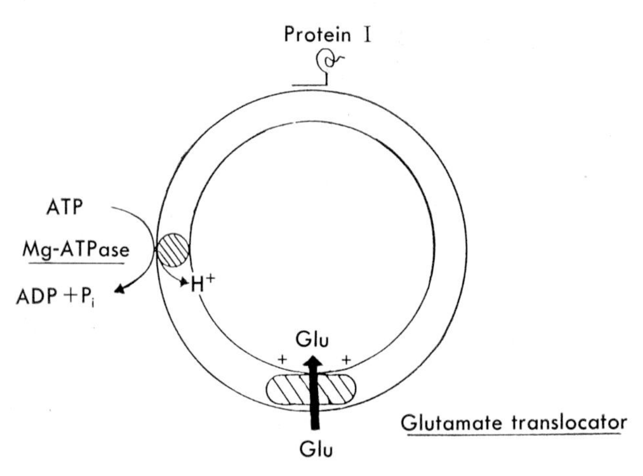

Fig. 4. Proposed mechanism for glutamate uptake into synaptic vesicles.

the glutamate uptake into the synaptic vesicle may also be driven by electrochemical gradients, a mechanism similar to that of catecholamine uptake into storage granules (Fig. 4). Namely, the Mg-ATPase in the synaptic vesicle translocates the protons produced upon its action on ATP, which results in the generation of a membrane potential, positive inside. This membrane potential provides a driving force for the glutamate uptake. When permeant anions are present at relatively high concentrations in the extravesicular medium, the membrane potential induces anion influx and thereby leads to lowering the intravesicular pH. The pH gradient thus formed also constitutes a driving force for the glutamate uptake. The relative contribution of the membrane potential and pH gradient, however, remains to be investigated.

It is unlikely that the phosphorylation of Protein I is involved in the glutamate uptake or its regulation, since the anti-Protein I IgG able to block both the cAMP-dependent and calcium-dependent phosphorylations of Protein I had no direct effect on the ATP-dependent glutamate uptake, nor did the immunoprecipitation of synaptic vesicles result in an increase in the specific uptake activity for glutamate in parallel to the increase in the specific content of Protein I. The possibility remains open, however, that Protein I may be involved in the regulation of the release of glutamate from the synaptic vesicles or in the regulation of glutamate metabolism.

VII. CONCLUSION

Certain synaptic vesicles in the central nervous system are endowed with an energy-dependent uptake system specific for L-glutamate, distinct from the sodium-dependent uptake systems present in the plasma membranes. The ATP-dependent vesicular uptake of glutamate appears to be driven by electrochemical proton gradients generated by a Mg-ATPase in the vesicles, a mechanism similar to that of catecholamine uptake into the storage granules of adrenal chromaffin cells. These observations favor the notion that synaptic vesicles are involved in the synaptic transmission of glutamate. It is suggested that the presence of the specific vesicular glutamate uptake system in a nerve terminal would initially determine the fate of glutamate as a neurotransmitter.

SUMMARY

Although there is growing evidence to support the notion that L-glutamate functions as a major excitatory neurotransmitter in the mammalian central nervous system, the role of synaptic vesicles in the glutamate neurotransmission remains unclear. We have recently isolated those synaptic vesicles which contain Protein I (recently referred to as Synapsin I), a neuronal, synaptic vesicle-specific phosphoprotein, using highly purified anti-Protein I IgG. These vesicles were found to take up L-glutamate specifically in an ATP- and temperature-dependent, but sodium-independent, manner. The ATP-dependent glutamate uptake system in the synaptic vesicle membranes is distinct from the sodium-dependent glutamate uptake system(s) known to be present in the plasma membranes of glia cells and nerve terminals; the former, which has a lower affinity for L-glutamate than the latter, does not recognize aspartate, whereas the latter transports aspartate as well as glutamate. The ATP-dependent vesicular uptake of glutamate requires ATP hydrolysis and appears to be driven by electrochemical proton gradients generated by a Mg-ATPase in the synaptic vesicle membrane, a mechanism similar to that for catecholamine uptake into the storage granules of chromaffin cells. These results support the notion that synaptic vesicles are involved in the synaptic transmission of glutamate.

Acknowledgments
This research was supported in part by NIH grant NS 1513. I thank Ms. Mary Roth for excellent assistance in the preparation of the manuscript.

REFERENCES

1 Abdul-Ghani, A.-S., Coutinho-Netto, J., and Bradford, H.F. (1981). In *Glutamate: Transmitter in the Central Nervous System*, ed. Roberts, P.J., Storm-Mathisen, J., and Johnston, G.A.R., pp. 155–203. Chicester, New York, Brisbane, Toronto: John Wiley & Sons.
2 Aswad, D.W. and Deight, E.A. (1983). *J. Neurochem.* **40**, 1718–1726.
3 Bashford, C.L., Cassey, R.P., Radda, G.K., and Ritchie, G.A. (1976). *Neuroscience* **1**, 399–412.
4 Bloom, F.E., Ueda, T., Battenberg, E., and Greengard, P. (1979). *Proc. Natl. Acad. Sci. U.S.* **76**, 5982–5986.

5 Burke, B.E. and DeLorenzo, R.J. (1982). *J. Neurochem.* **38**, 1205–1218.
6 Carty, S.E., Johnson, R.G., and Scarpa, A. (1981). *J. Biol. Chem.* **256**, 11244–11250.
7 Chujo, T., Yamada, Y., and Yamamoto, C. (1975). *Exp. Brain Res.* **23**, 293–300.
8 Cotman, C.W., Foster, A., and Lanthorn, T. (1981). In *Glutamate as a Neurotransmitter*, ed. DiChiara, G. and Gessa, G.L., pp. 1–27. New York: Raven Press.
9 Crawford, I.L. and Connor, J.D. (1973). *Nature* **244**, 442–443.
10 Curtis, D.R., Phillis, J.W., and Watkins, J.C. (1960). *J. Physiol.* **150**, 656–682.
11 Curtis, D.R. and Johnston, G.A.R. (1974). *Ergebnisse der Physiologie* **69**, 97–188.
12 De Belleroche, J.S. and Bradford, H.F. (1973). *J. Neurochem.* **21**, 441–451.
13 De Belleroche, J.S. and Bradford, H.F. (1977). *J. Neurochem.* **29**, 335–343.
14 De Camilli, P., Cameron, R., and Greengard, P. (1983). *J. Cell Biol.* **96**, 1337–1354.
15 De Camilli, P., Harris, S.M., Huttner, W.B., and Greengard, P. (1983). *J. Cell Biol.* **96**, 1355–1373.
16 Dolphin, A.C. and Greengard, P. (1981). *J. Neurochem.* **36**, 1627–1631.
17 Dolphin, A.C. and Greengard, P. (1981). *Nature* **289**, 76–79.
18 Goelz, S.E., Nestler, E.J., Chehrazi, B., and Greengard, P. (1981). *Proc. Natl. Acad. Sci. U.S.* **78**, 2130–2134.
19 Gordon, R.D. and Balázs, R. (1983). *J. Neurochem.* **40**, 1090–1099.
20 Greengard, P. (1976). *Nature* **260**, 101–108.
21 Greengard, P. (1982). *Harvey Lectures*, Series 75, pp. 277–331. New York: Academic Press.
22 Hackett, J.T., Hou, S.-M., and Cochran, S.L. (1979). *Brain Res.* **170**, 377–380.
23 Holz, R.W. (1978). *Proc. Natl. Acad. Sci. U.S.* **75**, 5190–5194.
24 Huttner, W.B., De Gennaro, L.J., and Greengard, P. (1980). *J. Biol. Chem.* **256**, 1482–1488.
25 Huttner, W.B., Schiebler, W., Greengard, P., and De Camilli, P. (1983). *J. Cell Biol.* **96**, 1374–1388.
26 Johnson, J.L. (1972). *Brain Res.* **37**, 1–19.
27 Johnson, R.G., Pfister, D., Carty, S.E., and Scarpa, A. (1979). *J. Biol. Chem.* **254**, 10963–10972.
28 Johnson, R.G. and Scarpa, A. (1979). *J. Biol. Chem.* **254**, 3750–3760.
29 Johnston, G.A.R. (1979). In *Glutamic Acid: Advances in Biochemistry and Physiology*, ed. Filer, L.J., Garattini, S., Kare, M.R., Reynolds, W.A., and Wurtman, R.J., pp. 177–185. New York: Raven Press.
30 Kennedy, M.B. (1983). *Annu. Rev. Neurosci.* **6**, 1–42.
31 Kirshner, N. (1962). *J. Biol. Chem.* **237**, 2311–2317.
32 Kontro, P., Marnela, K.-M., and Oja, S.S. (1980). *Brain Res.* **184**, 129–141.
33 Krnjević, K. and Phillis, J.W. (1963). *J. Physiol.* **165**, 274–304.
34 Krnjević, K. (1974). *Physiol. Rev.* **54**, 418–540.
35 Krogsgaard-Larsen, P. and Honoré, T. (1983). *TIPS* **4**, 31–33.
36 Krueger, B.K., Forn, J., and Greengard, P. (1977). *J. Biol. Chem.* **252**, 2764–2773.
37 Logan, W.J. and Snyder, S.H. (1972). *Brain Res.* **42**, 413–431.
38 McLennan, H. (1981). In *Glutamate as a Neurotransmitter*, ed. DiChiara, G. and Gessa, G.L., pp. 253–262. New York: Raven Press.
39 Naito, S. and Ueda, T. (1981). *J. Biol. Chem.* **256**, 10657–10663.
40 Naito, S. and Ueda, T. (1983). *J. Biol. Chem.* **258**, 696–699.
40a Naito, S. and Ueda, T. (1984). *J. Neurochem.*, in press.
41 Naito, S. and Ueda, T. Unpublished results.
42 Nestler, E.J. and Greengard, P. (1980). *Proc. Natl. Acad. Sci. U.S.* **77**, 7479–7483.

43 Nicoll, R.A. and Alger, B.E. (1981). *Science* **212**, 957–959.
44 Puil, E. (1981). *Brain Res. Rev.* **3**, 229–322.
45 Rassin, D.K. (1972). *J. Neurochem.* **19**, 139–148.
46 Roberts, P.J. (1981). In *Glutamate: Transmitter in the Central Nervous System*, ed. Roberts, P.J., Storm-Mathisen, J., and Johnston, G.A.R., pp. 35–54. Chichester, New York, Brisbane, Toronto: John Wiley & Sons.
47 Sandval, M.E. and Cotman, C.W. (1978). *Neuroscience* **3**, 199–206.
48 Sawada, S., Takeda, S., and Yamamoto, C. (1983). *Brain Res.* **267**, 156–160.
49 Schousboe, A. and Hertz, L. (1981). In *Glutamate as a Neurotransmitter*, ed. DiChiara, G. and Gessa, G.L., pp. 103–113. New York: Raven Press.
50 Schulman, H. and Greengard, P. (1978). *Nature* **271**, 478–479.
51 Sonnhof, U. and Bührle, C. (1981). In *Glutamate as a Neurotransmitter*, ed. DiChiara, G. and Gessa, G.L., pp. 195–204. New York: Raven Press.
52 Storm-Mathisen, J. (1976). In *GABA in Nervous System Function*, ed. Roberts, E., Chase, T.N., and Tower, D.B., pp. 149–168. New York: Raven Press.
53 Storm-Mathisen, J. (1981). In *Glutamate as a Neurotransmitter*, ed. DiChiara, G. and Gessa, G.L., pp. 43–55. New York: Raven Press.
54 Storm-Mathisen, J., Leknes, A.K., Bore, A.T., Vaaland, J.L., Edminson, P., Haug, F.-M.Š., and Ottersen, O.P. (1983). *Nature* **301**, 517–520.
55 Ueda, T., Maeno, H., and Greengard, P. (1973). *J. Biol. Chem.* **248**, 8295–8305.
56 Ueda, T. and Greengard, P. (1977). *J. Biol. Chem.* **252**, 5155–5163.
57 Ueda, T., Greengard, P., Berzins, K., Cohen, R.S., Blomberg, F., Grab, D.J., and Siekevitz, P. (1979). *J. Cell Biol.* **83**, 308–319.
58 Ueda, T. (1981). *J. Neurochem.* **36**, 297–300.
59 Ueda, T. and Naito, S. (1982). *Prog. Brain Res.* **56**, 87–103.
60 Ueda, T. and Greengard, P. Unpublished results.
61 Watkins, J.C. and Evans, R.H. (1981). *Annu. Rev. Pharmacol. Toxicol.* **21**, 165–204.
62 Zisapel, N. and Zurgil, N. (1979). *Brain Res.* **178**, 297–310.

SUBJECT INDEX

AUTHOR INDEX